Key Concepts STEM

SCIENCE TECHNOLOGY ENGINEERING MATH

ENGINEERING AND TECHNOLOGY

BUILDINGS AND STRUCTURES

General Editor: Anne O'Daly

BROWN BEAR BOOKS

Published by Brown Bear Books Ltd
4877 N. Circulo Bujia, Tucson, AZ 85718
USA
and
Studio G14, Regent Studios, 1 Thane Villas,
London N7 7PH, UK

Cataloging-in-Publication data available upon request.

ISBN 978-1-83572-061-5 (ALB)
ISBN 978-1-83572-067-7 (paperback)
ISBN 978-1-83572-073-8 (ebook)

Children's Publisher: Anne O'Daly
Design Manager: Keith Davis
Picture Manager: Sophie Mortimer
Consultant: Chris Woodford

Manufactured in the United States of America
CPSIA compliance information: Batch#AG/5664

Picture Credits
Front Cover: Shutterstock: Kirill Neiezhmakov. **Interior: iStock:** 19t, 25t, 25b, 57t, Ahrys_Art 12b, 60, Aksenovko 34t, aluxum 31, anthurren 37t, avstrallavasin 20l, Bim 35t, 44, S Borisov 40–41, Chris Boswell 32, brytta 12t, dageldog 19b, digital visions 11t, Paula French 40l, fuchs-photography 5t, gerenme 52, Avier Gill 16, Chris Hepburn 43t, JurgaR 36, Dmitriy Kazitsyn 34b, Livingimages 10l, luoman 43b, 61t, Monkey Business Images 50–51t, northlightimages 42, Allja Orbon 1, 29b, Sean Pavone 24t, 45b, ra-phots 7, Frederico Rostagno 28, Siegfried Schnepf 41, SergIV 58b, Daniel Taeger 37b, TerryJ 17l, tirc83 14b, Toby Alan 6b, Tpuerzer 27b, tpuerzer 27b, valleybol63 30–31; **Library of Congress:** 22r, 45t; **Photos.com:** 8t, 11b, 46, 47, 51bl, 53t, 54t; **Public Domain:** Averette 27t, Justine 35b, Roman Bonnefoy/romanceor 14t; **Shutterstock:** 501room 23, Achim Baque 8b, Justin Black 57b, chuyuss 28–29t, Mikhail Dudarev 9, Alexey Fedorenko 17r, Fiwsamatcha 22l, hainaul Hoto 18, jejim 33, Emin Kuliyev 20r, Sergiy kuzmin 53b, Lawrence Wee 24b, V Levi 48t, LNP Images 59, maziarz 4, Nick Poon 38, Howard Sandler 6t, SeventyFour 51r, Kaca Skokanova 48b, George SV 54b, Tivanova 56b, Tupungato 10r, Lawrence Wee 24b, Gary Yim 26; **Thinkstock:** Ablestock 30b, Hemera 5b, 56t; **Wellcome Images:** 58t.

STEM

Key Concepts in STEM: Engineering and Technology describes the scientific principles and feats of engineering that have shaped the world we live in. Since earliest times, people have created tools and machines to provide shelter, warmth, security, and protection from disease. This book looks at the technology and engineering used to create buildings and structures. It shows the development of building materials and design, and identifies the advances that have made it possible to build ever bigger structures. Explanatory diagrams and informative photographs explain how STEM (science, technology, engineering, and math) has transformed our lives and continues to influence our world today.

Publisher's note to educators and parents: Our editors have carefully reviewed the web sites included at the end of this book to ensure that they are suitable for students. Many web sites change frequently, however, and we cannot guarantee that a site's future contents will continue to meet our high standards of quality and educational value. Be advised that students should be closely supervised whenever they access the Internet.

Contents

TRADITIONAL STRUCTURES

Since the earliest times, people have constructed their buildings from materials available in their local environment. Until the invention of engines around 200 years ago, building materials were made and lifted using muscle power, both human and animal.

Today most buildings are made from bricks, steel, and concrete, with glass windows and shingled roofs. However, the earliest humans built temporary dwellings out of easily available materials such as grass, mud, and sticks using primitive tools. In some places humans lived in caves and didn't have to build shelters at all, while others simply lived under the open sky.

Most ancient peoples had no fixed homes. They were "hunter-gatherers," meaning they would travel from place to place, moving either when the seasons changed or when the animals they hunted moved to fresh pasture. Some of these tribes lived in tents, which they took with them whenever they moved.

There is a theory that the continent of America was first populated by people who walked across the Bering Strait separating Asia from Alaska when a land bridge emerged there during the last Ice Age. This is evident

Traditional building techniques worked well, and many old buildings, such as St. Paul's Cathedral in London, which was completed in 1710, are still standing. In 2000, a new bridge was built to connect the south of the city to the cathedral. It looks very different from the cathedral, but balances forces using strong, durable materials in exactly the same way.

The tents once used by many Native American groups are called tipis. These animal skin tents had holes at the top to allow smoke from fires inside the tipi to escape.

not only in the genetic characteristics of the Native American people, but also in comparisons of their dwellings. The typical dwelling of many Asian peoples is called a *yurt*, *jirga*, or *ger*, which is similar to a type of tent used by some Native Americans.

After the agricultural revolution from about 10,000 BCE, people in many parts of the world settled into more permanent dwellings. The wooden frames that had supported tents became more robust, and the animal hides and woven fabrics that had previously been used as coverings were replaced by sturdier, heavier materials made by mixing wood, grass, and mud. This change gave the structures a new strength and durability. Eventually, the wooden frameworks became stronger as larger dwellings were constructed.

SOCIETY AND INVENTIONS

Mobile Homes

While some nomadic people traditionally built new shelters from local materials wherever they went, other groups took their homes with them. The shelters of these people had to be reliable, weatherproof, mobile, and versatile. In order to meet these needs, a variety of basic tent designs were created.

Some of these styles have not changed for many hundreds of years. The Bedouin people of the Middle East and North Africa, for example, live in the same style of tent as their ancestors. Their tents are made from woven cloth that is pulled tight over poles, with only one wall and a roof to protect the occupants from the wind. The Mongolian *ger* (or *yurt*, below), on the other hand, is more like a portable house than a tent. The walls are made from criss-crossed fences, while poles stretching over the top form the roof. This frame is then covered by a thick blanket of felt that keeps the heat from the fire inside the home.

Everything needed to make a family's ger*—and the beds and furniture inside—are transported on dozens of camels, yaks, and horses. Everything can be packed away into tight bundles—except the* ger*'s solid wooden front door.*

Ancient Building Materials

In different parts of the world people had different materials available to them. To a great extent this defined the type of buildings that they constructed.

In the marshes of southern Iraq, for example, reed rushes and mud from the wetlands were used to make long pillars that were inserted into holes dug in the ground in two parallel lines. The tops of the pillars were then bent over and fastened together with the opposite pillar, making a row of arches. These arches were attached together by thin bundles of reeds, and reed matting then covered the entire building. These techniques are 6,000 years old and are still used today.

In places such as Scandinavia, the Swiss Alps, and the Himalayas, where timber has generally been plentiful, people have tended to make their houses from wood. Tall, straight-trunked trees such as pine, fir, and beech are particularly suited to this use. The log cabins of these areas are often made by flattening two sides of the long trunks and cutting the ends so that they fit together.

This longhouse made by the Native American people of eastern Canada has a frame of wood and was covered in skins in cold weather. The building could house up to 20 families.

MUD AS A BUILDING MATERIAL

Mud has been used for building all over the world. Mud was used by the Pueblo people of North America and also across the Atlantic in Mali, West Africa, where many mud buildings still stand. Made well, mud houses can be strong and also very beautiful. The great ziggurats (raised platforms) of ancient Mesopotamia were made from mud bricks. Despite being about 4,000 years old, the ruins of these mud constructions can still be seen.

The Great Mosque at Djenne, in Mali, West Africa, is made from mud and wood and is around 800 years old, although it has been rebuilt several times.

THE IGLOO

The Inuit, who live in the Arctic, still use snow to make temporary shelters, called igloos, in the same way they have for centuries. Igloos are made from big carved blocks of frozen snow and are formed in the shape of a dome, a design that offers great natural strength. Loose snow is then packed over the top, filling in gaps.

When the dome has been made, the builder goes inside, tightly sealing the igloo from the cold air outside. A lamp is then lit. The hot air, having no way to escape, begins to melt the blocks of snow. Then, when cold air is allowed in again, the melted snow quickly freezes. This process cements the blocks together and forms a smooth surface of ice on the inside of the igloo. The structure may be further strengthened by sleet, which freezes as it runs down the sides of the building. Igloos can be linked to one another via roofed passages, forming community dwellings of between three and five rooms capable of housing 15 to 20 people and keeping them warm at night.

Igloos are temporary shelters, built while Inuit people are hunting for seals far from home in the summer. Blocks of clear ice are fitted between the slabs of white snow to make windows.

Houses are still built completely out of timber in many parts of the world today.

Building in Stone

Buildings constructed from stone are strong because it is such a hard and durable material. Some of the earliest stone buildings date back as far as 2770 BCE—meaning they have survived for almost 5,000 years. One civilization that used stone to spectacular effect were the people of ancient Egypt. In order to extract the stone from a quarry, the Egyptians used pickaxes to chip away at the rock until five sides of a rectangular block

WATTLE AND DAUB

One of the oldest building materials is wattle and daub. This material has been found in some of the oldest settlements know, such as Çatalhöyük, a 9,000-year-old ruined city in central Turkey. Wattle and daub is a composite material, making use of two very different substances to create a sturdy structure. The wattle is a fence made by weaving thin branches around timber supports. This inner structure is then filled in and covered by the daub, a soft material made from mud, straw, horsehair, and even dung.

were exposed. They then drilled holes along the sixth face and filled them with wooden wedges. Doused with water, the wedges expanded, breaking the block away from the stone face. Masons then used mallets and chisels to shape the block into a smooth rectangle, checking their work with straight edges, set squares, and plummets (pieces of lead attached to a line used to check the alignment of vertical surfaces). Such simple tools remained in use all over the world until the Industrial Revolution in the 18th century.

Toombs and Temples

Stone was quite costly to quarry and maneuver, so it was mainly used for state or royal buildings. The most famous ancient structure made from stone is the Great Pyramid of the pharaoh Cheops (or Khufu) at

The Lycian people cut elaborate tombs out of the rocky cliffs of western Turkey about 2,400 years ago.

Giza, Egypt. Built over 4,500 years ago, it reached an incredible height of 481 ft (147 m), and was the tallest building in the world for almost 4,000 years. Similar pyramids can be seen in Central America. They were built from around 900 BCE by the Olmec, Toltec, and Mayan civilizations. Many of these temples have been found in the middle of thick rain forests. Like their Egyptian counterparts, the craftsmen used very basic tools to build these incredible structures.

The stonemasons of ancient Greece were equally highly skilled. Greek builders fitted blocks of stone together on site after trimming them so that they would slot into one another. Any gaps between the stones were filled with smaller pieces of stone.

INCA WALLS

The Incas of South America built strong stone walls (below, in Cuzco, Peru). The walls were dry—the stones were not glued together with cement. Instead, each stone was cut to fit together so precisely that you could not fit a knife blade between them. This was all the more amazing since the Incas only had cutting tools made from soft metals, such as bronze and gold, and stone axes.

In Africa, the ancient walled city of Great Zimbabwe (in modern-day Zimbabwe) was built from stone between around 1200 CE and 1450. At the height of its success, the city housed a population of nearly 20,000 people amd was spread over an area of 1779 acres (720 hectares). The inhabitants lived in thatched huts built from a mixture of clay and granite gravel.

By looking at the remains of the walls of this city, it is possible to see how the builders learned over time how to construct dry stone walls that were taller and stronger. The earliest walls are built unevenly, using different shaped blocks of stone. Examples of the later stonework have a better finish, and the stones were cut to uniform shapes and sizes to make a stronger structure.

THE GREAT PYRAMIDS

The main stone used for the construction of the Egyptian pyramids at Giza was limestone, with more beautiful granite used to provide a finishing layer. Limestone was readily available locally, but the nearest site from which granite could be quarried was Aswan, about 800 miles (1,300 km) away. Archaeologists are unsure as to how the stone was carried so far. One theory suggests that it could have been brought along the Nile River in barges. The barges probably brought the stone as close as possible, and then teams of people and animals moved the huge stone blocks small distances at a time, eventually reaching the building site. The next question is how these huge weights were transported to the top of the pyramid. Historians believe that the Egyptian builders made ramps from rubble, sand, or broken mud brick, and pushed the stones up the ramp using human and animal power. The angle of the ramp could not have been more than 10 degrees. If it was any steeper, humans and animals would not have been strong enough to get the stones up the slope. Historians think the ramps at Ghiza (below) were constructed in a spiral around the pyramids, and that they were taken down after the last stones had been laid.

Arguably, the ancient world's most impressive stone structure is the Great Wall of China, which was built to prevent nomads from raiding settled agricultural areas. Giving an exact date to the structure's origins is difficult, but the first attempts to build a unified wall date back to the 3rd century BCE. In the east, the core of the wall was built from rubble covered with an outer skin of stone. In this part of China stone was readily available. In the west, where stone was hard to find, the wall was made from a mixture of local soil and water. This was packed between two wooden barriers and then left to dry hard.

Bricks and Concrete

Bricks have been used for thousands of years. They were important in ancient Mesopotamia, while the earliest Greek temples were made from mud bricks and had thatched roofs. Some of the earliest and simpler Egyptian tombs were made from mud bricks. Because

USING CONCRETE

The ancient Romans used a mortar composed of sand, lime, and, water to glue the bricks of their buildings in place. However, sometime in the 2nd century BCE, they added a new ingredient to the mixture. This was *pulvis puteoli*, or *pozzolana*, a kind of ash that was found on the slopes of Mount Vesuvius, a volcano near Naples, Italy. When this was added to the mortar, it produced concrete, a material that became very strong and highly durable when dry. In fact, concrete was so strong that the Romans soon began to use it to form the interior walls, using brickwork simply as an outer shell. Eventually, some buildings were made entirely from concrete. The wet concrete was poured into a wooden framework that was removed as soon as the concrete had hardened.

After the fall of the Roman Empire the technique of making concrete was lost for many centuries. It was revived in Britain in the 18th century. Portland cement—a high-quality concrete extracted from quarries on Britain's south coast—was patented by the engineer Joseph Aspdin (1778–1855) in 1824.

Concrete slabs and walls are made by pouring the liquid mixture into wooden structures called forms. The concrete is smoothed over and allowed to dry hard before the forms are removed.

The Pantheon in Rome was made from concrete in 126 CE. The building's 42 foot (43.3 m) dome is the largest dome ever built from non-reinforced concrete.

they are held together with clay, which is found in most soils, bricks were available almost anywhere. They were also easier and more economical to use than stone.

To make bricks, clay or mud would be molded into a brick shape and left to dry in the sun. People later discovered that if bricks were heated in a kiln (very hot oven) to a greater temperature, they would be stronger and more durable. The size and shape of bricks also improved. Initially, bricks were modeled by hand into the shape of a loaf of bread, until builders discovered that it was easier to use a mold to make them into smaller, neater, more uniform shapes that could be baked more efficiently in the kiln.

As this close up of the Colosseum, the great arena in Rome, shows, the biggest Roman buildings were constructed from small bricks and then covered with a facing of stone slabs.

The Great Wall of China is 13,171 miles (21,196 km) of walls and trenches that connect natural features such as hills and rivers into a defensive barrier stretching across China.

Whereas the Byzantine and later European builders used brick itself as a decorative surface, the Romans usually covered, or faced, their brick buildings with stone, marble, or, more frequently, with concrete. Strong and versatile, concrete was an important feature of Roman architecture from about 150 BCE onward.

Arch, Vault, and Dome

A major development that allowed stone to be used to span large areas was the arch. Although the Egyptians and the Greeks had both used the arch, it was the Romans who

Mud bricks can be laid out to dry in the sunshine. They become hard after several days, even without the use of a kiln.

were responsible for the most important development of arch building. Romans arches built using stone or concrete were able to stretch much larger spans without the limitations imposed by the columns and flat beams that had been the major technology in use within the ancient Greek world.

One of the most impressive surviving Roman arches is the aqueduct at the Pont du Gard in Nîmes, present-day France. Built over 2,000 years ago, the Pont du Gard has three tiers of arches. The Colosseum in Rome, Italy, is another awe-inspiring example of the use of the arch. Constructed in around 75 CE, this amphitheater consisted of four tiers. The first three tiers were made from individual arches, to support the tall building.

THE POST-AND-LINTEL SYSTEM

One of the oldest building construction methods was the post-and-lintel system. It used a lintel, or beam, resting on a set of two posts, or columns, set into holes or foundations in the ground. One of the best known post-and-lintel constructions is Stonehenge (right). This 4,000-year-old giant stone circle was built on a high plain in southwest England. The post-and-lintel system was developed by the ancient Greeks.

The system has its limitations. The material used in the lintels has to be strong when placed under tension (when stretched), while that used in the posts has to be strong under compression (when squeezed). Stone cracks under tension, and stone lintels spanning long distances, will collapse. Arches, however, convert the tension into compression, which stone resists very well.

CRANES

The use of cranes is first mentioned in the work of the Roman architect Vitruvius. He wrote an instruction manual for builders in the first century BCE. The cranes of Vitruvius's day tended to be basic affairs, consisting simply of a long pole with a pulley (a wheel over which the rope passes) on the end of it. The pole was held in place by ropes. There was a winch at one end of the main rope. The pulley and the winch allowed large weights to be hoisted with comparatively little human effort. The Roman device was improved in the 15th century when the derrick crane (below) was developed by the architects of Renaissance Italy. The pulley was at the end of a movable arm known as a jib. This made the crane a far more flexible and effective machine. In some cases the crane would be powered by a treadmill. A crane powered by steam engines was invented in 1805, by Scottish engineer John Rennie the Elder (1761–1821). Hydraulic power came 41 years later, when British manufacturer William Armstrong (1810–1900) converted a steam-powered crane at Newcastle docks (in the north of England).

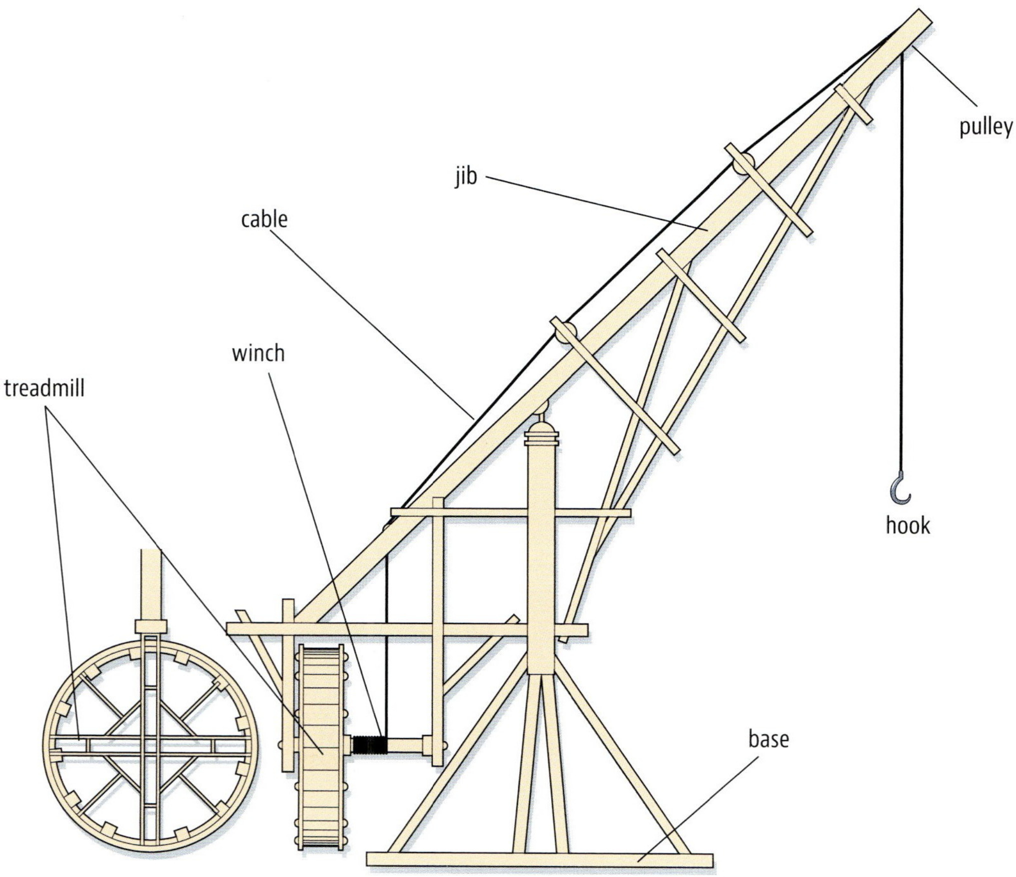

The principles behind the arch extend to the vault. A vault is basically an arch that has been lengthened to form a tunnel, and makes a solid ceiling. The vault can cover far greater amounts of space than traditional post-and-lintel structures. It was used in Mesopotamia (modern-day Iraq) as early as 3000 BCE and was further developed by the Romans. However, these simple barrel vaults either needed substantial buttresses (supports built against or projecting from a wall) or very thick walls to support their weight. The problem was partially overcome in the first century BCE with the introduction of the groin vault. Vaults were made to cross at right angles, and the edge formed at the point where the vaults crossed was known as the groin.

Heavy buttressing at the intersection allowed thinner walls to be used. The art of the vault developed slowly over the centuries and eventually reached its high point in the

The Church of Saint-Séverin in Paris has a high-vaulted ceiling held up by stone ribs that extend out from the wall columns.

SCAFFOLDING

While ancient Egyptian builders worked on tall buildings from specially constructed ramps, their Roman counterparts used scaffolding. This was a wooden framework constructed around the building from which the workmen could carry out their tasks. While little is known of the exact nature of Roman scaffolding, it is likely that it closely resembled the type used in medieval Europe. This was made from wooden poles lashed together with hemp ropes. The taller the structure, the thicker the poles. Platforms were made out of wicker, although these were replaced by wooden planks when heavy mechanical saws were introduced in the 15th century.

Modern scaffold is made from steel poles, locked together by clamps. In Asia, wooden scaffolds are still often used, even for tall buildings.

THE ARCH

The arch is an ingenious construction design based on mathematical principles. Arches work because the wedge-shaped blocks (*voussoirs*) that form the section at the top of the arch push the weight from above them down and out to the sides of the opening, allowing the weight to be carried by the two supporting posts (imposts). However, the force is not simply directed vertically downward: there is also a side thrust. Reinforcement is needed to stop the bottom of the arch continuing to move outward.

Because an arch cannot support itself during its construction, it needs to be built onto a temporary support. The temporary support, known as centering, is usually made from timber and is itself supported from underneath. The *voussoirs* are laid at either end on to the imposts and are built up until they almost meet in the middle, at which point the keystone is laid. The keystone is the central *voussoir* that holds the whole arch together.

In the 12th century, European architects discovered that the Roman semicircle (**1**) was not the most efficient form of arch. They found that the greater the height (or rise) of the arch as opposed to its span, the less side-thrust was exerted at its base. This discovery led to the Gothic arches (**2**) that created the churches built in late-medieval Europe.

1. ROMAN SEMICIRCULAR ARCH

2. GOTHIC ARCH

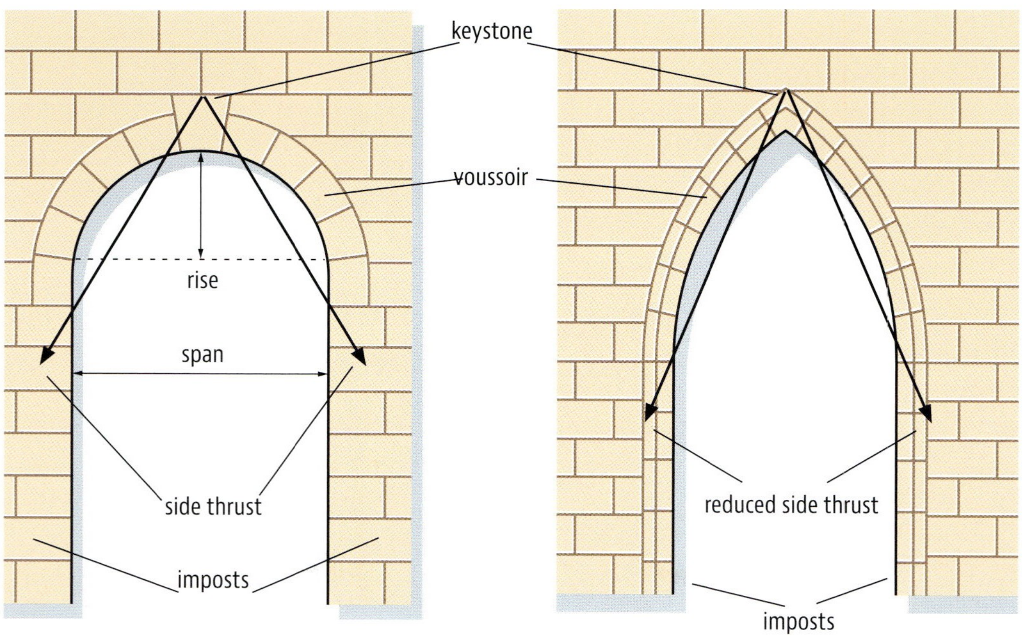

The tall towers of the cathedral in the city of Léon in Spain are connected by buttresses to provide extra support.

Gothic cathedrals of 16th-century Europe, when highly complex systems of interlocking vaults, arches, and flying buttresses (half-arches used as supports) were used to spectacular effect.

Improving the Dome

Another Roman structure that enjoyed a revival in medieval Europe was the dome. In principle, the dome worked like a circular arch. Because of the weight involved, the supporting walls had to be very thick to counteract the problem of side thrust. For example, the Pantheon in Greece, built in the 2nd century CE, needed walls 20 ft (6 m) thick

FACTS AND FIGURES

- The main section of the Great Wall of China is over 2,000 miles (3,200 km) long. The oldest section dates to at least 220 BCE.
- The oldest tent ever found is thought to be over 40,000 years old. It was discovered in Russia. Mammoth bones were used as the supports, and mammoth skin was the covering.
- The world's largest igloo is the Ice Hotel in Jukkasjärvi, Sweden. Rebuilt every year, the resort covers 4,500 square yards (3,500 sq. m) and has a church and a movie theater.
- The 481 ft (147 m) tall Great Pyramid at Giza, Egypt, contains approximately 2 million stones. Each block weighs an average of 2.5 tons (2.3 metric tons).

to support its 142 ft (45 m) span. In 1432, the Italian architect Filippo Brunelleschi (1377–1446) attempted to overcome the problem by placing ribs around the base of the dome of the cathedral in Florence. The dome of St. Paul's Cathedral, by the English architect Christopher Wren (1632–1723) and completed in 1710, is embraced by chains for exactly the same reason.

Apart from the development of better domes, however, the world saw surprisingly few advances in building techniques between the fall of the Roman Empire and the Industrial Revolution some 1,300 years later, when the discovery of new methods of making iron and steel would revolutionize the building industry. Even as late as the 18th century most of the techniques used by the world's builders would have been familiar to their ancestors who lived hundreds, and in some instances thousands, of years earlier.

The Dome of the Rock in Jerusalem was built in the 7th century. Its gold dome is constructed from wood on an octagonal base.

SOCIETY AND INVENTIONS

Chimneys

The Romans developed sophisticated methods of heating buildings—most notably the use of hot-air channels concealed in walls and under floors—but these techniques were lost with the collapse of the Roman Empire.

In early medieval Europe most buildings were based around large, communal rooms heated by a central fire. The smoke from the fire escaped through a hole in the roof. This was not a very satisfactory arrangement: the hole let in wind and rain, while considerable smoke stayed in the room.

These problems were solved by the arrival of the brick fireplace and chimney sometime between the 12th and 14th centuries. Most of the smoke was channeled up the chimney and out into the open air. The invention had a major social effect, allowing smaller rooms to be heated, and causing a shift away from large halls and communal lifestyles.

Hampton Court Palace, built by England's King Henry VIII 500 years ago, has 241 chimneys, many highly ornamented.

MODERN BUILDINGS

The first cities were built thousands of years ago, by the ancient civilizations of Egypt and Mesopotamia. In the last 200 years cities have grown to a large size and density—with buildings that have grown to match.

The change from an agricultural society to a mainly industrial one is known as the Industrial Revolution. It started first in Britain, around 1760, and spread rapidly to Europe. By the 1850s much of the Western world had become industrialized, but the same process

Burj Khalifa, Dubai, is the tallest building in the world at 2,717 ft (828 m). It takes one elevator just 60 seconds to reach the observation deck on the 124th floor from ground level.

LONDON'S CRYSTAL PALACE

Constructed to the design of the botanist and gardener Sir Joseph Paxton (1801–1865), the Crystal Palace was effectively a gigantic iron and glass greenhouse that was built for London's Great Exhibition of 1851. This huge structure—1,848 ft (563 m) long and covering an area of 991,857 square feet (92,243 sq. m)—used highly innovative construction techniques, which would have been impossible before the Industrial Revolution. The building was made entirely of standardized, interchangeable parts, the vast majority of which were mass-produced. The iron rods that formed the frame of the building were simply bolted together on site. The construction was highly organized and took only nine months. The palace was specifically designed so that it could be dismantled and reassembled elsewhere, and it was successfully moved to the suburb of Sydenham before being destroyed by fire in 1936.

is still taking place in other parts of the world today. It was the Industrial Revolution that first prompted the large-scale migration of people from the country to the cities to work in the new factories. This meant that cities had to be able to house large numbers of people; it was the dawn of the urban world. For all its chaos and complexity, the modern urban world has been made possible by a surprisingly small number of inventions, most notably the use of metal-framed buildings and reinforced concrete.

The Age of Iron

The Industrial Revolution brought about many dramatic changes, not least of which was the widespread use of iron in buildings. One of the first demonstrations of this was the iron bridge, constructed in 1779 at Coalbrookdale in the county of Shropshire, England. Iron was not only cheaper and stronger than timber or stone, but it could also be produced on a

THE EIFFEL TOWER

The most famous example of an iron-framed structure is the Eiffel Tower, produced for the Paris Exhibition of 1889. It was named for the French engineer Gustave Eiffel (1832–1923), who also designed the gigantic steel framework that supports the Statue of Liberty in New York.

The Eiffel Tower was the tallest building in the world until 1930. Its height is 1,063 ft (324 m).

The town of Ironbridge, which grew up around the pioneering all-metal iron bridge built near the Coalbrookdale mines, is known as the "birthplace of the Industrial Revolution."

SOCIETY AND INVENTIONS

Ancient Skyscrapers

Sana'a, the capital of Yemen, has the oldest apartment buildings in the world. These ancient skyscrapers are 1,400 years old and have several stories, despite being made from clay bricks. Just as in modern cities, people had to build upward to make the most of space in the crowded center of the city.

large scale. Steel, which was even stronger, was mass-produced from 1856 onward, following the invention of the Bessemer process, which mass-produced steel from molten pig (crudely processed) iron.

The Bessemer converter was named for English engineer Henry Bessemer (1813–1898) who invented it. It wasn't long before iron and steel was being used in buildings. They were used to construct the huge Crystal Palace in London, and soon engineers began to have even bigger ideas.

Balloon Framing

By the middle of the 19th century iron-framed construction started to make an impact in the United States, initially in Chicago. It was the second time the city had been at the forefront of building innovation. From around 1830, Chicago had expanded rapidly thanks to the invention of balloon framing by the American engineer G.W. Snow. This method constructed buildings on wooden frameworks: long vertical sticks ran from roof to floor, and boards or joists were nailed to them. Balloon framing cost half the amount of traditional carpentry and was much faster.

The Flatiron Building, constructed in New York City in 1902. One of the first skyscrapers, its name refers to its long, triangular shape.

Skyscrapers

Ironically, the success of balloon framing also led directly to its downfall. In 1871, a fire destroyed many of the wooden balloon-framed buildings in Chicago. As a result, a group of architects, known as the Chicago School, began making buildings with iron frames.

As cities started to grow, the price of land in their centers started to rise. It became important to make the maximum use of the space available. In theory, a building could always be made taller, but there were two practical difficulties with this. First, people could not be expected to walk up hundreds of stairs. Second, the weight of the masonry made it impossible to make a building above 16 stories.

The first of these problems was solved in 1854 when American inventor Elisha G. Otis (1811–1861) developed an elevator with an automatic safety device. The second problem

ELEVATORS

In an elevator, the passenger car is hauled up and down by a set of strong steel cables called hoisting ropes. These loop around a pulley wheel, or sheave, which is attached to an electric motor on the roof of the building. The weight of the car is balanced by a counterweight, which considerably reduces the power needed to lift the car.

Early elevators were rickety and dangerous. If the main cable broke, the elevator crashed to the floor. That changed in 1854, when Elisha G. Otis invented an elevator with a safety device. In the Otis elevator, the car traveled between guide rails that had toothed edges. If the cable broke, metal levers sprung outward from the car, locked into the teeth, and stopped the car from falling.

In 1857, Otis's company introduced the world's first ever passenger elevator at the New York china store of E. V. Haughwout. Powered by steam—electric elevators wouldn't arrive for another 25 years—Otis's elevator could transport 6 people at 40 ft (12 m) per minute. Modern-day elevators are capable of traveling at far greater speeds, up to 1,700 ft (510 m) per minute. In fact, they are theoretically capable of moving even more quickly, but it is thought that passengers would find the experience unnerving and unpleasant.

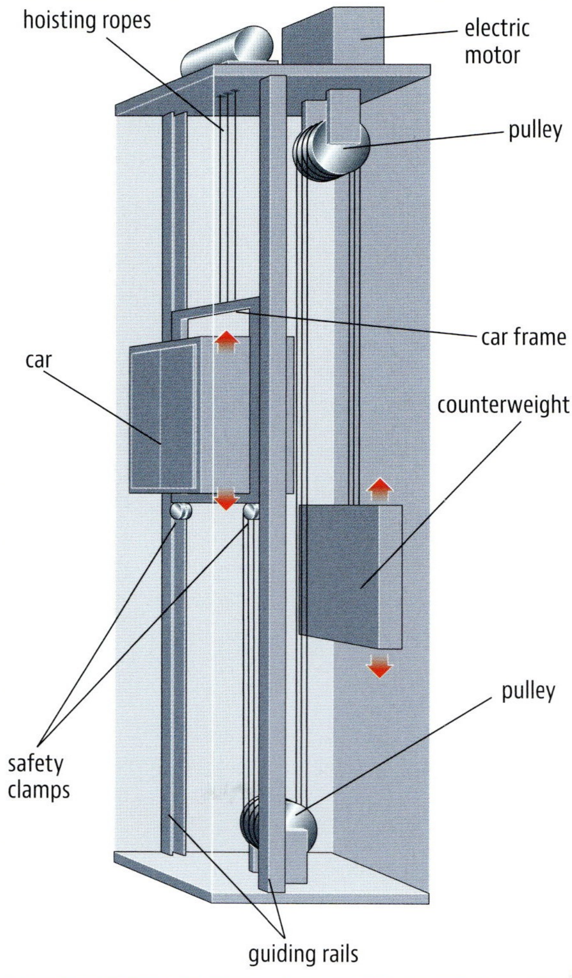

FORM AND FUNCTION

The new structural technologies that emerged in the second half of the 19th century in Chicago and France led to buildings that were increasingly functional. Outside, they were essentially steel skeletons covered with a skin of masonry, concrete, and glass. Inside, they still contained load-bearing walls, which helped support their weight. In many ways, the interiors were little different from the buildings of the past.

That changed with the French architect Auguste Perret (1874–1954), who discovered that a building with a strong steel framework no longer needed internal, load-bearing walls. His revolutionary apartment building at 25b rue Franklin, in Paris, did away with the internal walls to produce large areas of open space. It was a forerunner of the open-plan designs that are commonplace today.

Cleaning the windows on a building clad, or covered, in glass is a big job. Window cleaners often use mountaineering equipment to scale tall buildings.

Concrete can be surprisingly elegant, as Frank Lloyd Wright showed with these slender columns, supporting the ceiling of the Johnson Wax building he designed in 1936.

was solved by Chicago engineer William L. Jenney (1832–1907), who developed a strong skeleton of cast-iron columns and beams for his Home Life Insurance Building (completed in 1885). These two inventions removed the previous height limitation on buildings and led directly to that most essential feature of modern cities, the skyscraper. This new design completely revolutionized the architecture of modern city centers and changed civil engineering.

REINFORCED CONCRETE

While concrete has existed in some form since Roman times, it wasn't until the mid-19th century that the idea of reinforcing it with iron and steel was conceived. The first experiments in reinforced concrete were made by the French gardener Joseph Monier (1823–1906) who, in 1867, filed a patent for flowerpots that were reinforced by an iron cage. Monier's ideas were soon developed by the French builder François Hennebique (1842–1921), who began to use iron bars to strengthen masonry floors.

Hennebique realized that a straight iron bar would not reinforce concrete as efficiently as one that was bent in order to give support specifically to the areas where it was needed most. Concrete is strong in compression (when squeezed) but weak in tension (when it is stretched). Figure **1** shows a beam supported by three columns and the areas where the forces of tension and compression are greatest. The straight iron bar in figure **2** strengthens the concrete but does not support the areas where the tension is highest. However, changing the shape of the iron bar as shown in figure **3** delivers support to the areas where the concrete is under greatest stress.

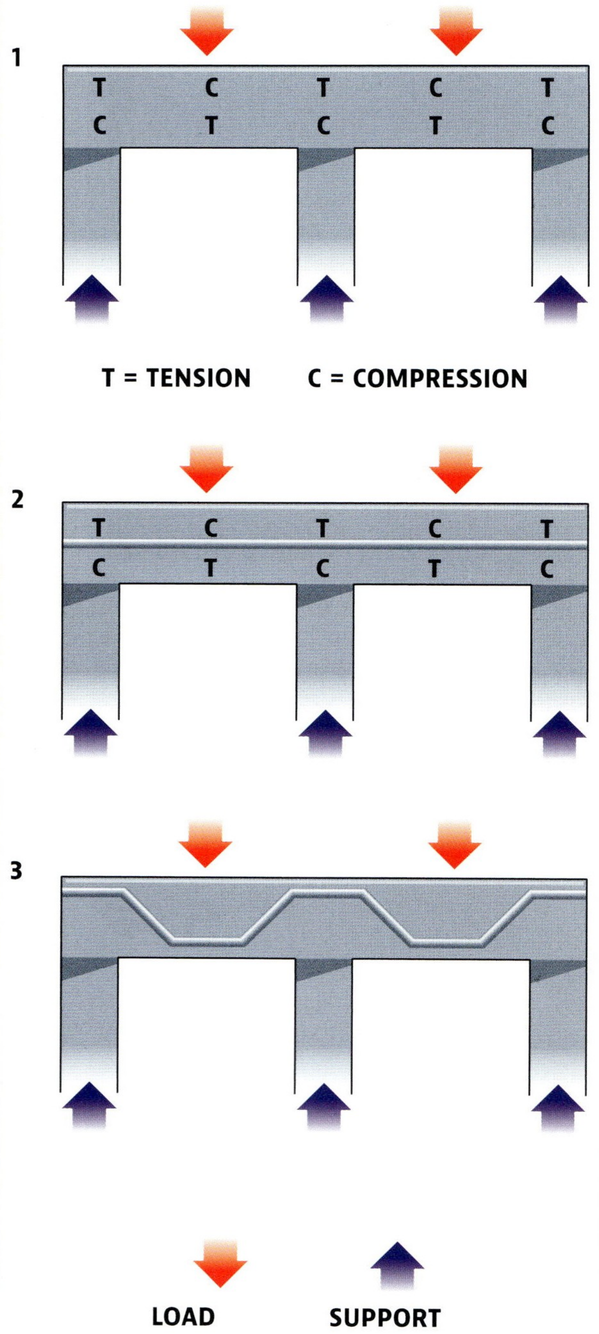

Steel reinforcements poking out from a column of concrete. If the metal corrodes inside the concrete, it is possible that the concrete will then crack.

The Gateway to the West arch in St. Louis has a core of reinforced concrete covered in steel plates.

Concrete and Steel

For all its innovation, William Jenney's first skyscraper still had stone walls. In 1892, French engineer François Hennebique patented reinforced concrete, which was a method of embedding steel bars inside concrete blocks to produce a material that was much stronger than either concrete or steel alone. Reinforced concrete literally changed the face of architecture and is used in almost every large building and structure today.

What most people regard as typical modern architecture developed at the Bauhaus school of design, which operated from 1919 until 1933 in the German cities of Weimar, Dessau, and Berlin. Under its directors, the German architects Walter Gropius (1883–1969) and Ludwig Mies van der Rohe (1886–1969), this hugely influential school pioneered a stripped-down, highly functional approach to architecture that was known as Modernism. It was inspired by a devotion to the power of machines and belief in a utopian vision of society.

The most influential modern architect of the time was the Swiss architect Charles-Edouard Jeanneret (1887–1965), mostly known as Le Corbusier. From 1923, when he published

AIR CONDITIONING

In the 20th century, air conditioning became a common feature of buildings. One of the key figures in the development of air conditioning was the American Willis H. Carrier (1876–1950).

Various types of air-conditioning units are in use today, but they all work from the same principles. Incoming air is cleaned, either by fiberglass wool filters or jets of water, before the temperature is adjusted. This is often done by passing the air over coils filled with either a refrigerant or hot water. Finally, in order to achieve the desired level of humidity, moisture is either added to or taken away from the air before a system of fans blows it back into the room.

Air-conditioning units, which can be seen in this side alley, pump heat and moisture away from the buildings.

his ideas in the book *Vers une Architecture* (*Toward a New Architecture*), Le Corbusier was one of the most innovative and controversial figures in architecture. He believed that a home should be a "machine for living in," a belief based on his admiration for automobiles and trans-Atlantic liners of the time, and he pioneered the concept of mass, low-cost housing. When most people think of Modernist architecture, they conjure up pictures of stark grids of concrete, steel, and glass. But the Modernist revolution led to some dazzling designs around the world. One of the best examples is the Trans World Airlines building at John F. Kennedy

Modernist buildings can be simple boxes that create large, open living spaces. Windows often make up an entire wall, leading to light, airy interiors that are connected to the outside world.

SOCIETY AND INVENTIONS

Designs for Living

The Unité d'Habitation in Marseille, France, was built between 1946 and 1952. It was a revolutionary experiment in social housing by Le Corbusier. Inside, this block of 337 apartments over 18 floors was like a small town housing 1,800 people. Wide corridors run all the way through, leading to split-level apartments that have their own staircases. The roof of the building includes a garden, swimming pool, and play area for children that even features caves and tunnels. The building also has shops, a school, and an open-air theater.

Le Corbusier's ideas were misinterpreted by other architects who tried to copy his ideas using small budgets. Sadly as a result, Le Corbusier is unfairly remembered as the architect whose work led to badly designed high-rise apartment buildings.

When properly designed and maintained, high-rise apartment blocks bring people together.

FRANK LLOYD WRIGHT

Although much of modern architecture can be traced back to the gridlike skyscrapers pioneered in Chicago from the 1850s, not all buildings developed along that route. The influential Chicago architect Frank Lloyd Wright (1867–1959) developed his own style of Organic Architecture. By using natural materials, his buildings seem to grow dramatically, yet harmoniously, from their environment. With their long horizontal planes close to the ground, they are instantly recognizable. His most famous building, Fallingwater, a large house built in Pennsylvania in 1937, is constructed above a waterfall, which is integral to the design.

International Airport in New York. Completed in 1962 and designed by the Finnish architect Eero Saarinen (1910–1961), its elegantly sweeping wings are a particularly imaginative use of reinforced concrete. Saarinen also designed the Gateway to the West arch in St. Louis, Missouri. Modernism has also led to the adoption of striking internal features in buildings, such as central atriums.

One of the most startling and original approaches to the problem of enclosing a space to make a building was proposed by the

The National Congress building in Brasilia features two half spheres. The left dome contains the Senate, while the lower House of Deputies sits in the right, upturned hemisphere.

An atrium runs down the center of a skyscraper hotel. This space provides light, but also acts as a chimney for rising warm air. There is often a pool at the base of an atrium to collect rainwater.

American architect and engineer Richard Buckminster Fuller (1895–1983). Fuller invented the geodesic dome, a huge sphere made up from hexagons. Fuller's biggest dome had a diameter of 384 ft (117 m) and was produced for the Union Tank Car Company in Baton Rouge, Florida, in 1958. He is perhaps best known as the designer of America's pavilion at the Expo '67 exhibition in Montreal, Canada. The pavilion was a huge 254 ft (75 m) dome covered with a plastic skin. The space frames used in public exhibition buildings and stadiums work in a similar way to Fuller's geodesic dome.

GEODESIC DOMES AND SPACE FRAMES

The weight of the materials in a traditional building, plus the weight of its occupants and contents, must be supported at key points in its structure. The load is supported by the external walls and some of the internal (load-bearing) walls. In a modern skyscraper the load is distributed evenly throughout the building's metal framework. Geodesic domes support the weight of a building or roof by distributing it evenly through a latticework of triangles and hexagons. Each strut in the framework carries the same load. Because there are so many struts, each one bears a relatively small amount of the load and can be made from lightweight material. Intermediate pillars or walls are unnecessary, so these structures are ideal for sports stadiums, museums, and exhibition halls requiring large open spaces.

This geodesic dome houses Science World in Vancouver, Canada. It was opened in 1989.

GREEN HOMES

Increasing concern for the environment has led to the construction of buildings designed specifically to create as little pollution as possible. Green buildings often get their power from solar panels on the roof or large areas of glass that trap the Sun's heat. Some green buildings are designed so that air circulates freely through the building without the need for air conditioning. Other types of green building are designed to blend into their environment, by being built partly underground or having their roofs covered with turf and other plants.

The two Bosco Verticale (Vertical Forest) apartment blocks in Milan, Italy, were completed in 2014. They were designed to increase housing in an already densely populated area while at the same time improving the air quality. Some 20,000 trees and perennial plants absorb the city's carbon dioxide and also protect the residents from the noise and dust of the busy streets below.

Shanghai in China is famous for its beautiful skyline. The city has some of the world's tallest buildiings.

Modern Cities

Modern cities are not simply collections of concrete skyscrapers. Since the time of the ancient Greeks and Romans, cities have been planned with varying degrees of success. The evolution of some of the greatest cities in the United States was planned in great detail: a plan for Washington, D.C., was worked out as early as 1791. In 1957, a new capital city for Brazil was planned. Brasilia, famous for its many unusual buildings, was the work of architect Oscar Niemeyer (1907–2012), one of the key figures in Modernism.

Environmental Planning

Modern ideas about city planning owe as much to environmentalism as to architecture. At the end of the 19th century English visionary Sir Ebenezer Howard (1850–1928) thought up the concept of the garden city as a response to the terrible living conditions that had been caused by the Industrial Revolution. The garden cities were carefully

zoned into business and residential areas with regions of permanent green belt to give an impression of the countryside. In the 1920s and 1930s these ideas were successfully copied elsewhere, for example in the American city of Greenbelt, Maryland.

Recognizing that cities have an enormous impact on the environment, city planners are now designing car-free streets with public buildings, such as hospitals and schools, as close as possible to transport links. Yet city planning always seems to be one step behind what people want. High-speed Internet connections have allowed people to work, study, and shop at home and so avoid the need to commute into a city at all. Just as the Industrial Revolution shaped today's cities, so the technological revolution of the 21st century will shape the cities of the future.

SOCIETY AND INVENTIONS

Planning Cities

Some of the world's oldest cities, such as London, England, have evolved haphazardly from many small villages and towns. By contrast Paris, France (right), was rebuilt according to a plan drawn up in the 1850s by Georges-Eugène Haussmann (1809–1891), although many of its buildings are now admired for their old-fashioned charm. In most cases, today's cities contain a mixture of modern planned precincts and older, historic districts that attract tourists.

Planning a city involves careful thought about where to place housing, offices, factories, recreation areas, railroads, subway systems, and highways. Typically, different activities are planned to take place in different areas, or zones. City plans include conserving historic buildings, regenerating areas that have become derelict, and ensuring the city and the people that live there have a minimum impact on the environment.

Twelve broad avenues in the center of Paris, France, spread out in a star from a central point known as Place de l'Etoile—meaning Square of the Star. These boulevards were planned to replace the narrow streets of the former city center.

CONTROLLING WATER

Water is vital for life and also important for transport and industry. Over many years, people have devised ways to control and use water effectively. These efforts have led to some of the largest engineering projects on Earth.

Ocean-going cargo ships pass through huge locks in the Panama Canal, a waterway through Central America that connects the Atlantic and Pacific Oceans.

The first major water-supply schemes, going back more than 5,000 years, were used for irrigating crops. For drinking water, people would have gone to the nearest clean stream or spring, or used specially dug wells that connected to underground water supplies. The construction of large-scale public supply systems began with the development of cities. Jerusalem, for example, had extensive underground water-storage facilities from around 1000 BCE.

The Roman Empire had the best known ancient water supply system. Many of the massive engineering works still survive to this day. Miles of aqueducts were constructed to

AQUEDUCTS

Aqueducts are long channels that transport water from high natural sources to lower-lying towns and cities. The first examples were built by the Minoans in ancient Crete and date back to 2000 BCE. Aqueducts rely on gravity to function and have to slope gradually downward throughout their length. Because of this, they were often forced to take fairly roundabout routes. The Aqua Marcia, which was built in 144 BCE and supplied water to Rome, stretched 57 miles (91 km), even though the water source was only 23 miles (37 km) away. It carried around 46 million gallons (171 million liters) of water each day.

The Pont du Gard aqueduct in France is 2,000 years old. It carried 44 million gallons (200 million liters) of water a day.

supply the fountains and bathhouses of ancient Rome. The aqueducts brought water from springs in the hills into the city. For crossing deep valleys, the Romans (and the Greeks before them) sometimes used inverted siphons—pipes that carry water down a gradient and up the other side. Siphons work if the pipes are strong enough, are always full of water, and are able to release the water at a lower level than it originally entered.

Various machines were used for lifting small amounts of water in the ancient world. These included the *shadoof*, a balanced pole with a counterweight for raising water out of ditches; the *saqiya*, a chain of pots worked by animal power; the Archimedes's screw, a screw-shaped device rotated by hand to lift water; and early forms of pump.

Cleaning up

The Roman water systems were not equaled in scale until the 18th and 19th centuries. At that time better water supplies were urgently needed to combat waterborne diseases in the new industrial cities of Europe, as well as to supply industry itself. Gravity-based systems for water distribution were supplemented by those in which steam-powered pumps propelled the water through enclosed water pipes or up into water towers, a transition that eventually led to the large-scale water supply systems of today.

WATER SYSTEM IN ISTANBUL

The Roman city of Constantinople, now Istanbul, Turkey, kept its water in cisterns, an underground storage system. The city has several hundred cisterns, many built in the 6th century by Emperor Justinian. The city's water came from a forest several miles to the north and was carried to the cisterns by a series of aqueducts. In 1555, after the city had transferred to Muslim rule, Sultan Suleiman built a new water supply system that channeled spring water to public fountains.

The Basilica Cistern is the largest in Istanbul. Its forest of columns created room for 100,000 tons (90,700 metric tonnes) of water.

Flood Control

Many civilizations have grown up around large rivers, yet most rivers inevitably flood from time to time. Sometimes the flood is predictable, as in the annual Nile floods that made ancient Egypt's land fertile. But unpredictable floods can cause disaster. For some rivers such as the Huang He (Yellow River) in China, flood-control measures go back thousands of years. Ways to protect against flooding include deepening and straightening the river channel, building levees (high, reinforced banks), and constructing extra overflow channels in cities.

The effects of flooding can be devastating for agriculture, industry, and people's personal lives. The town of Pacific Junction in Iowa (below) was submerged when a levee broke in March 2019.

Many large schemes to control rivers were introduced in the 20th century. Some of these schemes included the building of dams to trap floodwater, but rivers are complex systems and human intervention can sometimes have unwanted side effects. Flood defences can also give a false sense of security. Around 80 percent of New Orleans was flooded in 2005 after Hurricane Katrina struck the city, causing levees and flood walls to fail in more than 50 different places.

Dams

Dams have been built since ancient times. They were built mainly to store water or to divert it for irrigation. The earliest known dam was built at Kosheish on the Nile River in Egypt in around 2900 BCE and was used to supply water to the Egyptian city of Memphis.

The Three Gorges Dam on the Yangtze River in China is one of the largest engineering projects ever undertaken. The dam's reservoir is 410 miles (660 km) long.

TYPES OF DAMS

There are two types of dam: gravity dams and arch dams. Gravity dams are large embankments that rely on their size to withstand the weight of the water behind them. Arch dams are narrower and use their arch shape to transmit the force of the water to the side. They are built from concrete and are constructed in narrow gorges, where the surrounding rocks are strong.

Gravity dams are built from a mixture of earth, rock, or concrete. These dams are often porous and have a central core of a nonporous clay to hold back the water. A stone lid (known as rip-rap) is built on top to stop erosion from wave action.

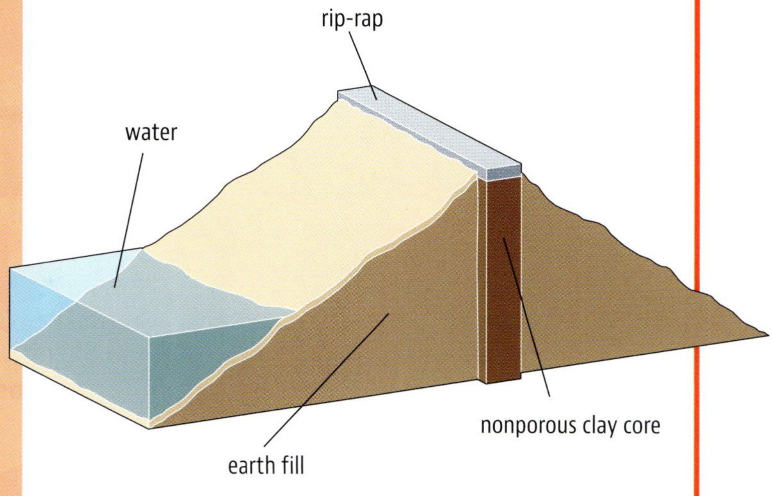

The largest gravity dams are embankments made from earth or other materials taken from the local area.

Unsurprisingly, many of the early developments in dam design came in the Middle East, because of its lack of regular rainfall. The Babylonians (from modern Iraq) and Assyrians built dams from around 700 BCE.

In the last 200 years considerable efforts have been made to use scientific principles to maximize the efficiency of dam design. In the middle of the 19th century engineers such as the Scotsman William Rankine (1820–1872) began to look at the forces exerted on dams

An excavated Roman latrine. In the Roman Empire, everyone used a communal toilet—there were no cubicles. There was also no toilet paper. Instead, everyone shared a sponge on a stick.

and ways that different-shaped dams resisted these forces. One shape that proved to be most effective was the arch. As dam design became more scientific, so larger dams could be built. In many cases a hydroelectric plant is built into the dam to harness the power of the water current.

WATER PURITY

The purity of water is important if people are to keep healthy. The Romans filtered water through sand, or boiled it. They used water from one source (such as spring water from the hills that arrived via aqueduct) to drink and used other water sources for washing clothes. They also left sediment to settle out of water before using it.

In many countries, water from a faucet is normally safe for washing and cooking but may require further purification before drinking. Filter jugs (below) are used to take out unwanted minerals and the chemicals used in reservoirs that can affect its taste.

Disposing of Sewage

Effective means of disposing of domestic sewage came much later in history than water supplies. Although ancient cities such as Mohenjo-Daro (in modern Pakistan) had elaborate drainage systems, in most cases they were aimed mainly at removing surface water. Until the 19th century human waste was often thrown untreated into the nearest river (as it still is in some countries today). The toilets that came into use from the 18th century onward often discharged into pits, or cesspools. The growth of large industrial cities made matters worse. In England in the 1850s, a cholera epidemic was traced to sewage contaminating public wells. When the

SEWAGE TREATMENT

In a modern urban sewerage plant the sewage goes through two or three stages before it is discharged back into rivers and waterways. In the primary treatment stage the sewage passes through a filter that removes the largest bits of waste matter. It next flows into a grit chamber, where substances such as sand settle to the bottom of the tank. The sewage then passes into a primary sedimentation tank where fine solids slowly sink to the bottom.

In the second stage the sewage is further purified by exposing it to the natural action of bacteria. One way to do this is to pass compressed air through the sewage (the so-called activated sludge process). The third treatment stages can include subjecting sewage to radiation treatment.

Greasy liquid floats to the surface of sedimentation tanks, where it is skimmed off by booms. Chemicals are added to make smaller particles clump together into larger units until they sink to the bottom of the tank.

importance of clean drinking water was recognized, some industrial countries then started to construct sewerage systems. At first, sewer pipes were connected to the existing drains that were used to collect storm water. This meant that untreated human waste passed out into nearby rivers or the sea. However, in the late 19th century specialist sewage-treatment plants were built.

The first sewage plants had tanks that allowed suspended material to settle out as sludge before the water was discharged into a river. This is called primary treatment. Modern sewage plants use more elaborate purification systems and the water is often recycled rather than being discharged.

A reconstruction of the water pump in London, England, where Doctor John Snow first realized that cholera was transmitted by water, and that clean water was critical to public health.

Protecting Against the Sea

Since ancient times, harbors have been provided with outer defenses (breakwaters) to shelter ships at anchor. Modern breakwaters often make use of huge interlocking shapes of concrete called tetrapods to absorb the force of the waves. Others are built as vertical walls to reflect rather than absorb the energy of the waves.

Similar defensive engineering projects are also applied to the coastline in general. River mouths are most vulnerable, being low-lying and often with large human populations. Storm surges—inrushes of the sea—can cause disastrous flooding. Coastal defences are particularly important today because of the rise in sea levels being caused by global warming.

FACTS AND FIGURES

- On average the world's city population consume 80–160 gallons (300–600 liters) of water per person every day. This contrasts with just 5–8 gallons (20–30 liters) per day in rural areas of less-developed countries.
- The amount of water on Earth is fixed and can never be used up. Scientists have estimated this to be about 333 million cubic miles (1,400 million cubic km). Around 97 percent of this water is in the world's oceans.
- At 1,388 million cubic feet (40 million cubic m), the world's largest dam is the Three Gorges Dam in China. The highest dam is the Jinping I Dam, at 1,001 ft (305 m), also in China.

The Thames Barrier in London, England, is a rotating dam that can block the Thames River and stop tidal surges flooding the city.

Global Concerns

Climate change and rapid population growth is putting great strain on the world's water resources. Scientists are looking at ever more ingenious ways of managing water.

Desalination plants, which remove the salt from sea water, have been used in California and the Middle East since the 1950s, but they are expensive to run unless cheap energy is available. Multipurpose water resource programs now try to combine irrigation, hydroelectric power, and flood control to maximize the potential benefits of these huge, costly projects.

SOCIETY AND INVENTIONS

The Delta Plan

In 1953, the delta of the Rhine River in the Netherlands was devastated by a major flood. Thousands of people died, and a vast amount of land was ruined by saltwater. The area soon became the subject of a major sea-defense project known as the Delta Plan. Large inlets of the sea were sealed off by dams, forming tideless lakes. As people became more aware of environmental issues, however, the original plans were changed. One major dam was replaced by a storm-surge barrier (right) that only operates when the tide is dangerously high. Otherwise the tides ebb and flow as normal and the environment is preserved.

Dutch flood barriers have protected the land, but the coastline is steadily eroding because the dams mean less silt is being dropped by the Rhine River as it empties into the sea.

BRIDGES AND TUNNELS

When people began to build roads and paths, they had to find ways to go over or through obstacles rather than travel around them. The answer to this was to plan and construct bridges and tunnels.

The Hong Kong–Zhuhai–Macau Bridge (HZMB) is 34 miles (55 km) long. It consists of bridges, tunnels, and artificial islands, and is the world's longest fixed sea crossing.

A bridge takes a road, called the deck, over the top of an obstacle such as a river. To do this, it has to have a special structure that will support the weight of the road and anyone on it and distribute the weight toward the points where the bridge meets the ground.

The most basic type of bridge is the beam bridge, formed naturally when a tree falls across a stream. The tree trunk takes the weight of anyone walking across it and distributes it to the riverbanks on either side. The distance between the two supporting banks is known as the span. Because it is unsupported in the middle, the trunk bends under the weight of someone standing in the

TYPES OF BRIDGE

All bridges are variations on four simple designs—the beam (or girder), arch, suspension, and cantilever. These four main types of bridge vary mainly in the way that they bear their load and then distribute it to their different types of support.

In a simple beam bridge (**1**) the weight is borne by the two supporting pillars and thrusts straight down. As the beam bends under the weight of its load, it is stretched. The material used for a beam bridge therefore has to be able to withstand great tension. Ancient beam bridges that were made from stone had limited spans because stone does not cope well with tension.

In order to overcome this, the Romans and other early civilizations began to use arches to build their bridges. In an arched bridge (**2**) the load exerts a sideways outward force, so arches were firmly buttressed at the side.

In a suspension bridge (**3**) the deck is held by cables suspended from towers. Some of the weight pushes downward through the foundations of the towers. The majority of the weight is held by the cables, which are anchored deep in the ground on both sides of the bridge.

The last type of bridge (**4**) is the cantilever. Here pairs of identically balanced sections are joined by a small central span. Any downward thrust through the pillars is counterbalanced by an equal and opposite upward thrust where the bridge meets the shore.

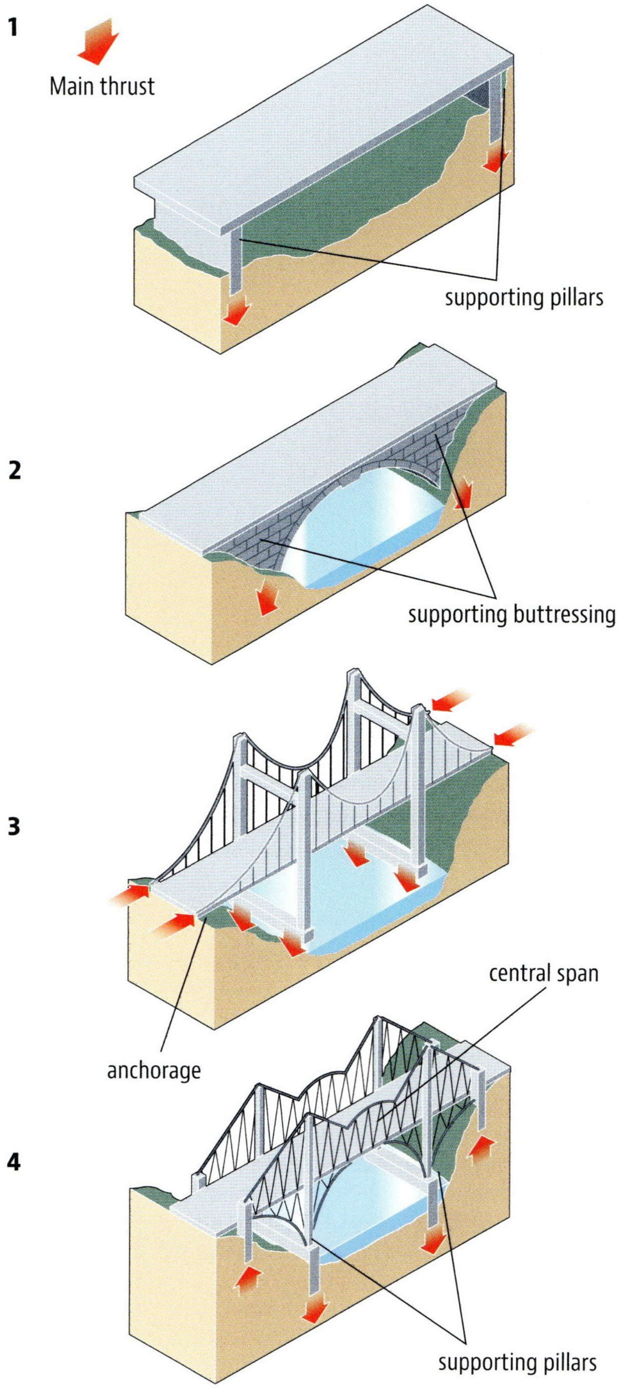

CLAPPER BRIDGES

The oldest surviving human-made bridges are variants on the beam bridge, known as the clapper bridge. Here, piles of flat stones are stacked on top of each other to form supporting piers, while larger, longer stones are laid across the top of them to form the beam. Clapper bridges were built by the Stone Age people of Western Europe as long ago as 2000 BCE. Many examples survive, and some are still in use today.

This 700-year-old clapper bridge was once a main route through Devon in southwestern England. It has two central piers, supporting three large stone slabs.

middle and puts the wood under stress—the top of the trunk is being compressed, while the bottom is being stretched. If the tree trunk is not strong enough, it will snap. This is a basic engineering problem for anyone building a bridge, and there are a number of different solutions.

A clapper bridge is a bridge with a beam built of stone, but the same method can be used to build wooden bridges. In prehistoric times people built trackways through marshes by hammering wooden piers down into a river bed and then laying beams across them. Another type of beam bridge has two parallel beams along each edge, with planks laid between them to create a wide deck.

Girder Bridges

A girder bridge has a beam made from either steel, concrete, or both. An advance on the girder bridge was the box girder, first used in a bridge by British railroad engineer Robert Stephenson (1803–1859). His Britannia Bridge—a railroad crossing over the Menai Straits in Wales, built between 1845 and 1850—consisted of two rectangular wrought iron tubes. The box shape made the tubes extremely strong. Stephenson had planned to give extra support with suspension cables, but when the final section of girder was in place, he realized that the bridge was strong enough without them.

The Rialto Bridge is a 420-year-old arch bridge that crosses the Grand Canal in Venice, Italy. The bridge has a covered walkway that is lined with shops.

Arched Bridges

Many of the oldest bridges still standing today use arches to support them. The arch works in just the same way as it does in a building, transferring its own weight and the weight of the deck down and to the sides. Arches avoid the bending forces that can cause beam or girder bridges to break because all the parts are compressed under their own weight. The only way an arch can fall down is if it is allowed to spread to the sides—a series of arches will push against each other where they meet, so this is only a problem at the points where the arches meet the banks (the abutments). Here, the bridge must have solid foundations.

For two thousand years the only arched bridges known in the West used semicircular arches. This was a problem because the deck had to run across the top of the arch, making it difficult to achieve a flat deck suitable for horse-drawn vehicles. The solution was to use a much smaller part of a circle for the shape of the arch. The Chinese were building segmental arched bridges from 600 CE, but

TRUSS BRIDGES

One successful variation on the girder bridge is the truss bridge. A triangular latticework of trusses is put up on either side of the deck. This stiffens the structure and makes it far less likely to break when subjected to heavy loads. The first truss bridge was built by Italian Andrea Palladio (1508–1580) and crossed Italy's Cismon River in 100 ft (33 m) spans.

Properly built, truss bridges are very stable. Most of the bridges that carried the railroad west across the United States had frameworks of wooden trusses. The first person to properly understand how they worked was the American engineer Squire Whipple (1804–1888). Now, truss bridges are made from iron and steel.

Trucks cross a truss bridge built in Narooma, New South Wales, Australia. The triangular structure of the girders gives such bridges great strength.

this technique did not reach the Western world until 1345, when the Ponte Vecchio ("Old Bridge") was built in Florence, Italy.

The arched bridge has remained popular to the present day. Modern arched bridges are usually built in one piece from steel or reinforced concrete.

Cantilever Bridges

The cantilever bridge was another design that originated in Asia. It consists of a long beam extending from a pier that is counterbalanced by another beam on the other side of the pier. The first cantilever bridges were made using planks of wood, each extending slightly farther out over the span. The cantilevers were joined in the middle by a short suspended section. This made the cantilever bridge a very stable design.

In the Western world the cantilever design was popular for railroad bridges, where the cantilever was built from iron or steel and balanced by an identical section extending to

FACTS AND FIGURES

- The bridge with the longest single span in the world is the Çanakkale Bridge in Turkey. The bridge's deck runs for 6,637 ft (2,023 m) between the supporting piles.
- The first suspension bridge to feature steel cables, rather than chains, was the Brooklyn Bridge in New York City. It had 1,200 miles (1,931 km) of steel wire, while 100,000 tons (91,000 metric tons) of masonry were needed to anchor the cables.
- One of the most ambitious bridges of recent years is the Danyang–Kunshan Grand Bridge on the high-speed railroad from Beijing to Shanghai. It opened in 2011. The length of the elevated deck is 102.4 miles (164.8 km), and the bridge is just one of 244 (plus 22 tunnels) on the railroad.

The Forth Bridge, built in 1890 to carry a railroad into Edinburgh, Scotland, is the world's second longest cantilever bridge after the Quebec Bridge in Canada.

The Clifton Suspension Bridge, spanning the Avon River near Bristol in England, was designed by Isambard Kingdom Brunel (1806–1859) and opened in 1864.

meet the land on the opposite side of the pier. Because all the weight was pushed down through the bridge's piers, it was a good design for carrying heavy vehicles such as trains. The first modern cantilever bridge was built across the Main River in Germany in 1867.

Suspension Bridges

Suspension bridges are another very old type of bridge, and again, no one knows who invented them. The oldest known examples are rope bridges from South America and southeast Asia. In these bridges two long ropes are slung across the gap to be bridged and securely fastened onto the land at either end. The deck is hung from these main ropes by shorter, vertical ones. Suspension bridges did not become popular in the Western world until after the Industrial Revolution in the 18th century. British engineer Thomas Telford (1757–1834) built the first modern suspension bridge across the Menai Straits in Wales in

CABLE-STAYED BRIDGES

The cable-stayed bridge, with its arrangement of overhead wires, looks like a suspension bridge. However, it works in a different way and is more closely related to girder and cantilever bridges. In a suspension bridge, the cables run freely across the tops of its towers and are only anchored at the ends, where the bridge meets the solid ground. In a cable-stayed bridge, however, the cables are attached directly to the towers, which support the bulk of the weight.

The cable-stayed bridge connecting Aracaju with Barra dos Coqueiros in Brazil was designed by Joao Alves.

Using modern materials, contemporary girder bridges are used as highways to span wide valleys and areas of woodland.

1826, using two huge towers to support the main cables. To counter the bridge's weight, the towers had to be anchored to the ground with more cables.

Suspension bridges were most popular in the United States. The engineer John Roebling (1806–1869) built a series of bridges using iron and steel cables. Roebling died before his most famous creation, New York's Brooklyn Bridge, could be finished, and the project was completed by his son Washington (1837–1926) in 1866. At this time the Brooklyn Bridge had the longest central span in the world at 1,595 ft (487 m). Another famous American suspension bridgeis the spectacular Golden Gate Bridge of San Francisco. Completed in 1937, the distinctive red bridge features a 4,200 ft (1,280 m) central span.

THE FIRST TUNNEL

The earliest known transport tunnel was a foot passageway built under the wide Euphrates River at Babylon (in modern-day Iraq). It was difficult to tunnel in the waterlogged riverbed, so the ancient engineers found a solution in diverting the river during the dry season, allowing the tunnel to be built. It was given a stone lining and then the water was allowed back.

Tunnels

Tunnels, like bridges, are designed to get around obstacles, but instead of going over them, they go under or through them. Different building methods are used for digging them and holding them up, depending on the type of ground. The ancient Egyptians hollowed out the tombs of their pharaohs using copper saws and reed drills, but this was only possible because Egyptian

The brick towers of Brooklyn Bridge (right) were constructed underwater on the riverbed. Workers entered the water inside pressurized workspaces called caissons.

SOCIETY AND INVENTIONS

Disasters

Bridge design has developed considerably over the centuries, with bridges becoming increasingly strong and durable and capable of covering longer and longer spans. Sometimes, however, bridge builders have overreached themselves, with disastrous consequences. When it opened in 1940, the Tacoma Narrows Bridge, which crossed Puget Sound in Washington State, was seen as a marvel of modern engineering, the third longest suspension bridge in the world. However, when motorists started to use it they found it alarmingly unstable and prone to swaying violently from side to side.

On November 7, just four months after it opened, the bridge collapsed in a light wind. When it was reconstructed, engineers were careful to widen the deck and stiffen it with a series of supporting trusses.

The collapse of the Tacoma Narrows Bridge was captured on film. No one was hurt in the disaster. Motorists abandoned their vehicles and fled moments before the bridge crashed down.

rock is soft and sandy. Once the cavern had been hollowed out, there was a danger that the pressure from the rock above it would make it collapse, so the cavern was immediately lined with stone blocks. These linings were formed by two uprights piers supporting a horizontal lintel, or beam, which held up the weight of the earth above it.

Some of the most impressive early tunnels were built in Persia (modern Iran) around 700 BCE and were used for irrigation and to supply water to towns. These tunnels could be many miles long and were built by digging vertical shafts into the ground and then tunneling along to join them together. The shafts also provided ventilation for the workers.

Railroads and Canals

From the mid-19th century, level ground was needed for railroads, which could not go up steep hills. Rail tunneling developed first in Britain and then in the United States, where the construction of the Hoosac Tunnel, linking

Tunnelers working on the Thames Tunnel under the Thames River worked inside a protective moveable cage or shield. The rocks and soil were carried away behind, while other workers lined the tunnel with bricks behind the shield.

ROMAN TUNNELING TECHNIQUES

Even though they had only primitive tools at their disposal, the Romans were able to construct impressive tunnels. One example, linking Naples and Pozzuoli and finished in 36 BCE, stretched for 4,800 ft (1,450 m). When they were tunneling through hard rock, the Romans employed a technique known as fire setting. This involved heating the rock face with fires, and then throwing cold water against it. The sudden cooling cracked the rock, allowing it to be dug out more easily. Tunneling through hard rock was more difficult than digging through soft ground but did away with the need for a lining. After the end of the Roman Empire, tunneling methods changed very little for over 1,000 years.

WATER TUNNELS

The ancient Greeks and Romans realized that tunnels could also be used for drainage. A tunnel under waterlogged ground would quickly fill with water. This had the effect of raining the land above it so the land could be reclaimed for building or farming.

Upstate New York and Boston, saw the first use of dynamite and steam- and air-powered mechanical drills. During the Industrial Revolution, canals also became a major transport network, and thousands of miles of new waterways were built in Europe and the United States. Bridges and tunnels were important for the canal network because the waterway had to be built absolutely level over long distances in order to stop the water flowing to one end. The first major canal tunnel was built for the Canal du Midi in France in the 1670s. This was also the first time that tunnelers had used gunpowder charges to blast away the rock. Explosives played an increasingly important role in tunneling projects from this time onward.

Tunneling Machines

The first tunnel under a major river was finally begun in 1825 by the French engineer Marc Brunel (1769–1849), who tunneled under the Thames River in London, England. Brunel realized the only way to keep the tunnel from flooding or collapsing was to line it with bricks as it was dug. He invented the first excavating machine—a circular framework pushed along the tunnel by hydraulic jacks. The frame was divided into cells, each of

London's Thames Tunnel, completed in 1842, was originally used by pedestrians and road traffic, before becoming part of the city's subway rail network.

which was occupied by a digger. As the diggers dug the tunnel, workers behind them lined it with bricks. By the time the framework was pushed forward again, a ring of brickwork was in place to keep the tunnel from collapsing. However, the project was still dangerous—the tunnel flooded several times when diggers hit pockets of water at the face and it was not completed until 1843.

Another way of building an underwater tunnel makes sections of tube on land and then sinks them into an underwater trench before filling in the seabed or riverbed on top. Despite being known since Brunel's time, this approach became popular from the 1950s.

A modern tunnel-boring machine (TBM) has dozens of hardened metal cutting teeth that grind away at the soil and rock as the borer's head spins around.

THE CHANNEL TUNNEL

One of the world's most spectacular tunnels is the 30 mile (50 km) Channel Tunnel, which joins Britain and France. Construction started in 1988 and the tunnel opened in 1994. The project involved 15,000 workers, working from both sides of the tunnel, and the final cost was approximately $15 billion.

Running 130 to 160 feet (40 to 50 m) beneath the seabed, the Channel Tunnel is in fact not one tunnel, but three—two main tunnels carrying railroad tracks and a third acting as a service tunnel for road vehicles to use. It was constructed using huge tunnel-boring machines. During the excavation, the machines removed eight million cubic yards (six million cubic m) of soil and rock. The completion of the tunnel has enabled people to travel directly from London to Paris, Brussels, and beyond by high-speed train rather than by boat.

Cars are transported through the Channel Tunnel on trains. The journey takes 35 minutes.

Today, many major tunnels are constructed using fully automated tunnel-boring machines (TBMs). These are often likened to giant mechanical earthworms. At their heads are huge circular cutting faces that rotate, chipping away at the rock or softer clay face. Behind the cutting heads robot arms line the tunnel with curved panels of concrete as the cutter burrows forward. The spoil (waste rock) is taken up to the surface by a series of conveyor belts. The whole machine is guided by a laser system that can detect and correct the slightest deviation from the planned path. TBMs move very slowly (around 30 times slower than a garden snail) so major tunnelling projects use two TBMs, which are calibrated to meet in the middle of the tunnel.

THE GREATHEAD SHIELD

For centuries the main difficulty facing people attempting to tunnel under rivers was the danger of water flooding the tunnel while it was being constructed. This problem was overcome by James Henry Greathead (1844–1896) when he invented the Greathead shield, a device that would not be bettered for 75 years. Greathead's machine was similar to earlier devices in that it featured a protective shield driven forward by hydraulic jacks. However, Greathead installed a compressed air chamber immediately behind the shield. The air pressure inside was great enough to push against the tunnel face and prevent water from entering the chamber. The miners working in the chamber were thereby protected from the dangers of accidental flooding. The compressed air chamber was separated from the rest of the tunnel by a concrete bulkhead. The only ways through were two airlocks, one for the tunnelers and one for the spoil coming out.

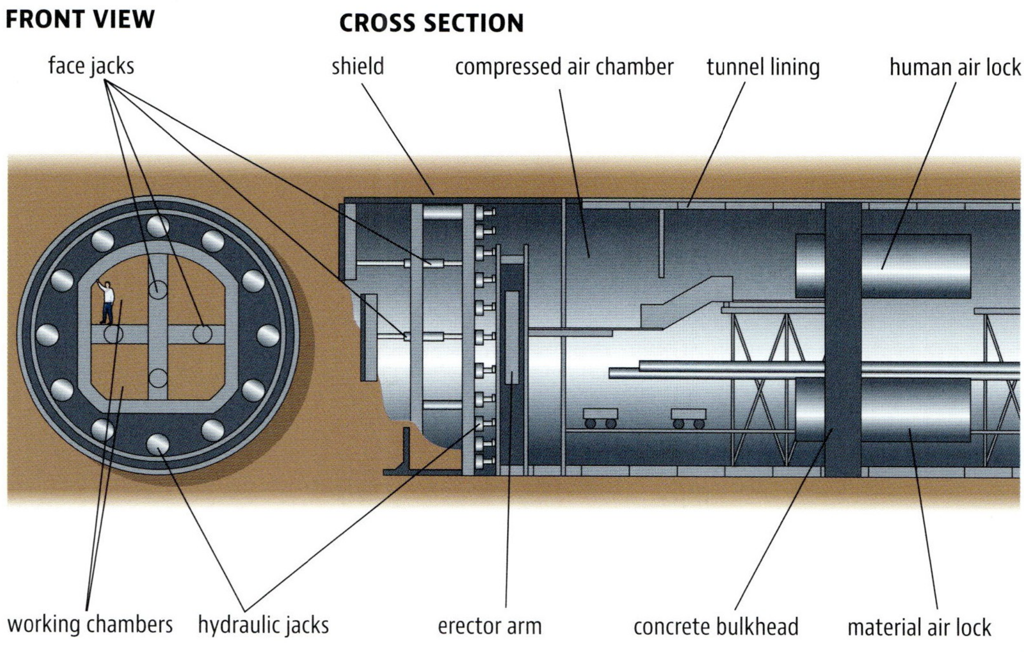

THE HOME

Home technology has made housework and cooking easier and faster, and in the modern home the kitchen and living areas are often together in one large room.

People discovered fire more than a million years ago. This changed the way they lived. They could now cook food and had a new source of light and heat. Technology in the home has continued to improve people's lives ever since.

For thousands of years, the main ways of cooking food were the open fire and an oven built of bricks. Stoves made from pig-iron began to appear in Europe and the United States after the Industrial Revolution (c.1760–1840). The stoves were fueled by coal or wood. However, from the mid-19th century, two new means of cooking began to revolutionize the way people prepared their food—natural gas and electricity.

Using Gas and Electricity

The first person to use gas to cook food was the German chemist Zachäus Andreas Winzler, who piped gas into the kitchen of his home in Austria in 1802. In 1824, the first gas cooking apparatus went on sale. Designed by Samuel Clegg, it was produced by the Aetna Ironworks in Liverpool, England, and consisted of a small tube with a number of holes in the side that could be placed horizontally to fry or vertically to roast food. The first gas stove was designed by the English entrepreneur James Sharp in 1826. It was made commercially available in 1834, and in the same year Sharp proved how efficient his stove was with a display of "Gastronomy" cooking, in which the stove was used to prepare a meal for 120 people.

From the 1870s the metal that stoves were made from began to be covered in a layer of enamel, which made them easier to clean and less likely to rust. In 1923, another step forward was made when the Regulo oven thermostat was introduced. This thermostat allowed the cook to set the oven at a specific temperature and eliminated much of the guesswork that had previously been necessary.

FACTS AND FIGURES

- The earliest known cookbook is the *De Re Coquinaria* (62 CE), by Apicius. He describes the feasts of the Roman Emperor Claudius I and his wives Messalina and Agrippina.
- In 1846, the American Nancy Johnson (1795–1890) came up with the idea of a freezing machine for making ice creams and sorbets. However, because the idea of women inventing objects was frowned on, Johnson registered the patent in the name of William Young.
- The effects of weightlessness in space mean that toilets have to be carefully designed. In the Space Shuttle, the toilet was flushed by a combination of water and air. It was also equipped with foot restraints and a seat belt.

Early examples of electric ovens were shown at an Electric Fair in London in 1891. In the same year a patent was taken out by the Carpenter Electric Heating Manufacturing Company of Minnesota for an electric heating unit made by enameling insulated wires onto cast iron plates. The system was improved by the English engineer R.E.B. Crompton (1845–1940), who used nickel alloys and a double

A cook prepares suet pudding—an old-fashioned dessert made from animal fat and flour that required long, slow cooking.

FOOD MIXERS AND KITCHEN GADGETS

The business of cutting, shredding, and mixing food can be a long process when done by hand. There have been many attempts to make these processes quicker in the kitchen. In the 1850s, the British company Nye and Lyon introduced the first domestic meat chopper, while in the United States, Enterprise developed machines as diverse as beef shavers, raisin-seeders, and sausage-stuffers. All these machines relied on hand operation but they helped to speed up the work.

One of the first electric machines was invented in 1918 by the American company Landers, Fray, and Clark. It was an electric food mixer that had two small beaters driven by electric motors. The biggest advance in kitchen automation, however, came in 1948, when English inventor Kenneth Wood designed the Kenwood Chef. It was an all-purpose machine fitted with a large motor that could drive beaters, choppers, squeezers, pasta-makers, and can-openers as required.

Blenders turn lumpy mixtures into smooth liquids. They can be used to make soups, sauces, milkshakes, and smoothies.

DISHWASHERS

In 1827, the French engineer Benoît Fourneyron (1802–1867) invented the first practical water turbine, which used flowing water to drive machinery. The principles he used were reversed for the production of the first dishwashers, developed in the United States in the mid-19th century. One such device, invented by Benjamin Howe in 1880, consisted of a frame that held the dishes in place while water was hand-cranked around the machine to clean the crockery. Just over a decade later, Josephine Cochrane showed off an improved version at the Chicago World's Fair of 1893. She did not design the machine to save herself washing up. Instead, its purpose was to prevent her servants from chipping fine china during cleaning! The first electric dishwasher appeared in 1912. Today's dishwashers dry the contents with heat after cleaning them so they are ready to use straight from the machine.

A modern dishwasher uses just over 3 gallons (14 liters) of water during a typical wash cycle.

layer of enamel to hold the wires more securely. Among the devices he sold were electric hotplates, as well as frying pans and saucepans that had electric wires embedded in their handles. Cooking with electricity was slow to catch on, however, since few houses had electricity until the early 20th century.

In 1924, the Belling company, which had been set up by an electrician who had worked for Crompton 12 years earlier, unveiled the first split-level electric stove, the Modernette. Five years later the company introduced its popular Baby Belling stove, with three hot rings and an oven beneath. Demand for electric stoves gradually increased after this period as more homes became connected to an electrical supply.

Keeping Food Cold

Keeping food fresh has always been difficult. As food gets warm, bacteria on the surface can feed and breed, turning the food moldy or making it rot. One way to keep food fresh is to keep it cool. Early larders had shelves made from materials such as slate and marble,

EARLY HOME FURNISHINGS

Some of the oldest surviving examples of furniture are found at Skara Brae on Orkney, off the coast of Scotland. This group of prehistoric stone houses, dating from 3180 BCE, was equipped with built-in benches and sleeping areas. More elaborate examples of furniture are found in the pyramid tombs of ancient Egypt. They include elaborately decorated stools, tables, chairs, and beds. The ancient Greeks and Romans were also very skilled furniture makers, developing folding chairs, cupboards, and chests. As homes have developed, separate rooms have gained specific uses and with them, specific types of furniture and accessories.

An electric stove based on the Baby Belling design.

which helped maintain a cooler atmosphere. In the late-19th century the first insulated ice-chests began to appear. These were usually lined with either zinc or slate. Ice was put inside the chest, and the food was placed on top.

Industrial refrigeration techniques were first developed in the 19th century to enable ships to carry meat and other perishables on long journeys from Europe across the Atlantic. In 1876, the German engineer Karl von Linde (1842–1934) designed the first domestic refrigerator. It used ammonia as a refrigerant, which was pumped around using a small steam pump.

In 1913, the Domelre (Domestic Electric Refrigerator) Company unveiled its first household refrigerator in Chicago. It proved a great success, and further developments were made by American inventor Nathaniel Wales, whose Kelvinator was first sold in 1916. The Frigidaire appeared one year later. Many of these early refrigerators were noisy because of the electric motors they used. However, in 1922 the Swedes Carl Munster and

COOKING WITH RADIO WAVES

In 1945, the American scientist Percy LeBaron Spencer (1894–1970) was standing in front of a magnetron, an electronic tube that emits short electromagnetic waves. He noticed that a chocolate bar in his pocket had melted and soon realized that the waves emitted had heated the bar. He patented the first microwave oven in 1945. A commercial version appeared in 1965.

When a microwave oven is turned on, the food inside is bombarded by high-frequency electromagnetic waves. The waves cause water molecules inside the food to vibrate, raising its temperature, which cooks it. A microwave oven cooks food in a fraction of the time that a conventional oven takes.

However, microwave ovens have drawbacks. For example, despite recent developments, they are still unable to brown or crisp foods as well as a conventional oven, so although food is made piping hot, some dishes just do not look or taste quite as appetizing.

Modern microwave ovens cook some foods in seconds or just a matter of minutes.

This ice-making machine from the 1860s used the compression and rapid expansion of ammonia gas to draw heat out of water, which turned it to ice.

Balzar von Platen built the first silent refrigerator that was able to operate in a domestic environment. This refrigerator was then mass-produced by the Electrolux company of Sweden in 1925.

Bathrooms

Baths of one form or another have existed for centuries. Early examples have been found in the Indus Valley (in modern-day Pakistan) that date back to 2500 BCE. However, despite this the bathroom has only become a commonplace part of the home in comparatively recent years. In Europe, for example, the bathroom practically

DOING THE LAUNDRY

For centuries, people have washed their clothes by hand in streams. In the absence of a natural running water supply, clothes were simply washed in wooden or iron tubs. Often a rod or paddle known as a dolly was used to squash and stir the clothes. This system was mechanized in 1858 by the American inventor Hamilton Smith. The first electric washing machine was invented in 1907 by Alva J. Fisher. A small motor was used to rotate the dolly. This design was improved in 1924, when a combined electric washer–spin drier was produced by the Savage Arms corporation of New York. However, it was a cumbersome machine. After washing, the drum—still full of wet clothes—had to be lifted and placed on another drive shaft. It was not until the 1950s that easy-to-use automatic machines were available.

disappeared after the fall of the Roman Empire and did not reappear in most houses until the late 19th century. It was then that systems were developed in which water was heated in a central boiler. One of the first of these was the Ewart Califont. Developed in 1899, it supplied hot water to every tap in the house from a boiler situated in the basement.

Bathtubs themselves were originally made from iron. However, as they got older they started to rust. The solution was to coat the iron in porcelain enamel. The first lightweight, enameled bathtubs were installed in Pullman railroad coaches in 1900.

Toilets, another essential item in modern bathrooms, have also been around for centuries. The first attempts to mechanize the toilet came in 1596, when a water closet designed by the English courtier Sir John Harrington (1561–1612) was installed at Richmond Palace in London. Harrington's water closet had a seat over a bowl and a pipe

HOW A REFRIGERATOR WORKS

Domestic electric refrigerators make use of the way gases behave when they expand quickly. Adding heat to a gas makes it expand, but a refrigerator makes a liquid, called a refrigerant, evaporate (become gas) and expand without adding heat. The expanding gas takes heat from its surroundings, so making them colder.

The refrigerator contains three main elements—an evaporator, a compressor, and a condenser. The refrigeration cycle begins when the refrigerant is pushed along a pipe and through a nozzle. This lowers the pressure of the refrigerant, turning it into a gas as it enters the evaporator. As the refrigerant vaporizes, it absorbs heat from the air inside the fridge, lowering the temperature. The cold gaseous refrigerant continues to travel around the maze of pipes that make up the evaporator until it reaches the compressor. The compressor is used to push the refrigerant into and out of the condenser. This increases the pressure on the gas, condensing it back into a liquid. As it does so, it gives out the heat it absorbed, which escapes out of the back of the refrigerator into the room at large. The cycle then begins again.

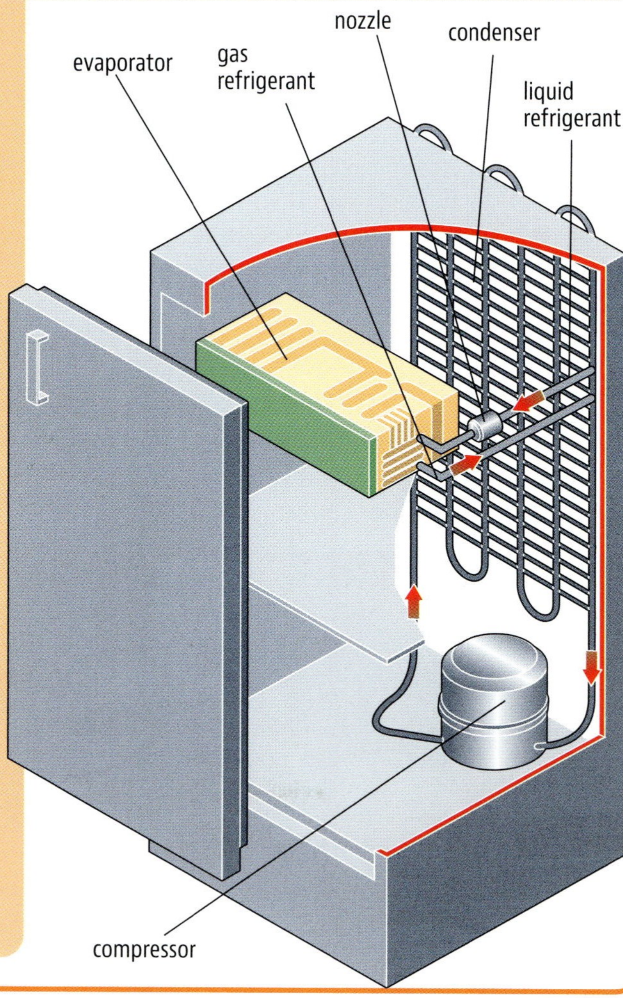

This old bathtub was carved from a solid piece of stone.

IN HOT WATER

The biggest obstacle to taking a regular bath was getting enough hot water. The simplest way to do this was to heat water in the kitchen and then carry it to wherever the bathtub was placed. In 1827, in order to overcome this inconvenience, the Thompson company produced a bath that contained its own heater. Water passed from a storage tank into a cylinder furnace and then into the bath. In 1868, Benjamin Maugham invented a special geyser that heated water for baths. It had a tube carrying water that spiraled through a cylinder containing gas burners.

that let water in to clean away the waste. Only two Harrington toilets were ever installed, and for 200 years nothing further was heard of the flush system. Then, in 1775, the English watchmaker Alexander Cummings patented the valve closet. It had two valves, one to let the water in and another that controlled the outlet at the bottom of the pan. Cummings's closet was modified by English cabinetmaker Joseph Bramah (1748–1814) and remained popular until the end of the 19th century.

These early toilets were made from metal. However, in 1870 the British potter Thomas Twyford (1849–1921) made the first all-ceramic toilet. It was Twyford who added the familiar S-bend outlet pipe, which provided a water-seal to prevent smells coming up from the drains. The system was further modified in 1889, when a British plumber named David Bostel invented the so-called washdown flush mechanism, the system that is still in widespread use today.

Before homes were fitted with running water and indoor bathrooms, people washed in their bedrooms using water poured from a jug into a large bowl, or basin.

Housework

American Hiram H. Herrick patented a carpet sweeper in 1858. This device was improved in 1876 by Melville R. Bissell (1843–1889), the owner of a porcelain shop who was allergic to the dust from the straw used to pack his pots. He developed a broom equipped with a cylindrical brush that pushed the dust into a container.

Litter is swept from city streets today by trucks equipped with rotary brooms and powerful vacuum cleaners.

SOCIETY AND INVENTIONS

Bathing

Public attitudes to bathing have changed wildly over the centuries and have varied considerably from culture to culture. Many ancient civilizations considered bathing to be a religious ritual. The importance of cleanliness to the Hindu lifestyle explains the existence of 4,500-year-old public baths in the Indus Valley in modern-day Pakistan. Ancient Rome was another place where bathing was common. Although few private houses had bathrooms, public baths were numerous and could hold as many as 1,600 people at a time. However, in Europe in the Middle Ages the practice of bathing declined drastically. In some countries even the wealthy did not bathe. For example, the 17th-century royal palace at Versailles, France, despite having elaborate fountains, did not have any bathrooms at all. Wealthy people used perfume to hide any bad smells. Bathing came back into fashion in Europe in the late-19th century. Today, attitudes vary widely from country to country. While in the Western world people use baths primarily to get clean, in Japan people are expected to wash before they get into the bath, which is used for relaxation.

The English city of Bath is named for its large Roman baths, which provided hot, warm, or cold baths from 60 CE.

BOSTEL'S TOILET

The design of the washdown toilet (below), originally the work of British plumber David Bostel, has been remarkably enduring. In Bostel's system an overhead cistern supplies the water that flushes the toilet bowl. When a chain or handle is pulled, a plunger is raised within a syphon unit. Water is then pushed through an inverted U-bend and down into the bowl, washing away the contents. When this happens, the water level inside the cistern drops, until a float, called the ball cock, falls to a certain level opening an inlet valve. As the water rises, so does the ball cock, which closes the valve when the cistern has refilled ready for the next flush.

In the 19th century, chimney sweeps were regular visitors to houses, clearing the soot from chimneys with long, flexible brooms.

In 1901, the English civil engineer Hubert Cecil Booth (1871–1955) designed the first vacuum cleaner. He had seen presentation cleaning machines that blew dust from the carpet and realized that dirt could be collected more effectively using suction. Booth's British Vacuum Cleaning Company used large horse-drawn vacuum cleaners that traveled from house to house and were operated by a team of uniformed employees. Booth's machine was cumbersome—it was powered by a heavy gas motor, while its suction hoses were 100 ft (30 m) long. These problems were overcome in 1907, when the American inventor Murray Spengler developed a much lighter machine that was powered by a small electric fan. However, Spengler was unable to market his invention and sold the rights to a leather manufacturer named William Hoover (1849–1932). The Hoover Model "O" was launched in 1908 and became an international success.

VACUUM CLEANERS

The first vacuum cleaners, developed in the early 20th century, blasted air down into the rug and tried to catch the dust as it flew up. Modern cleaners work the opposite way. Most use rotating brushes to loosen the dirt and a powerful fan to suck it up. After capturing the dirt, they return fairly clean air to the room.

Until the 1990s, most vacuums used large fabric or paper bags to catch their dust. They had a powerful electric motor that drove brushes at the front and a fan in the middle. The brushes spun at high speed, beating the dirt loose from the carpet, which the fan sucked up and blew into the dust bag at the back. The bag acted as a filter, trapping the dirt and allowing clean air to return to the room. As the bag slowly filled with dirt, the cleaner became less effective.

Since the 1990s, many people have switched from bag-type vacuums to more effective cyclonic cleaners. These still use rotating brushes to loosen dirt and powerful motors to suck it in. However, instead of using a filter bag, they spin the dusty air around a spiral path called a cyclone. Just as a washing machine dries clothes by spinning out water, so a cyclonic cleaner spins dirty air to separate out dust. This is trapped in a large, easy-to-empty plastic bin. The cleaned air passes through two very fine fabric filters so hardly any dust returns to the room. Unlike bag-style cleaners, cyclonic vacuums work just as well whether their dust bins are empty or full.

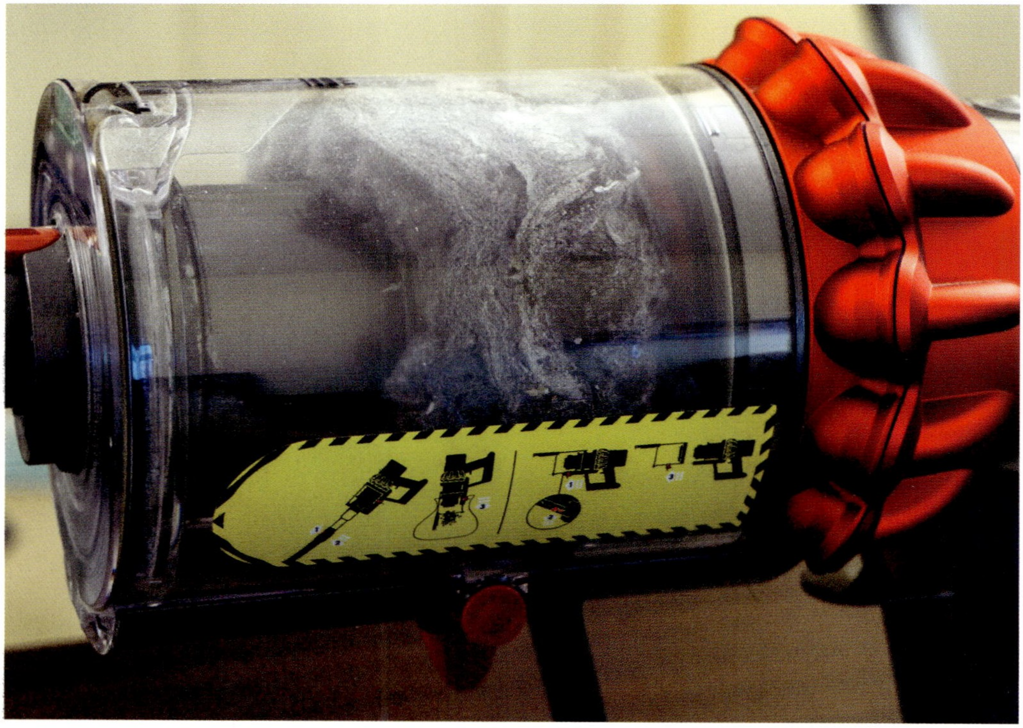

A cyclonic vacuum cleaner separates dust and dirt from the airflow. The spinning motion pushes the dirt to the outer walls of the chamber.

TIMELINE

1,800,000 BCE The earliest tools so far discovered were made around this time in the Olduvai Gorge, Tanzania, for cutting meat.

1,000,000 BCE Humans discover fire.

30,000 BCE The first baked clay objects are made.

10,000 BCE The first clay pots are made.

4000 BCE Glazes are used to make pottery watertight.

2900 BCE The earliest dam is built on the Nile River in Egypt.

2700s BCE Early stone buildings date from this period.

2560 BCE The Great Pyramid of Giza is built over a period of about 15 years. The pyramid is the tallest building in the world for more than 3,000 years, until Lincoln Cathedral is built in 1311 CE.

1500 BCE Iron smelting is first carried out on a large scale.

1000 BCE Underground water-storage facilities are built in Jerusalem.

600 BCE Chinese develop the blast furnace to make iron.

62 CE The earliest known cookbook is published.

800 The Chinese produce porcelain, a strong but light form of china.

1494 Monk John Cor distills whiskey for his abbey.

1712 Thomas Newcomen develops a steam engine for removing water from mines.

1779 The first iron bridge in the Western world is constructed near the Coalbrookdale mine fields in England.

1784 Henry Cort patents the puddling process for making wrought iron.

1794 Jacob Schweppe introduces the fizzy drink after Joseph Priestly discovers, in 1770, that carbon dioxide dissolved in water creates a refreshing drink.

1795 Nicolas Appert develops a new canning system for preserving food by heating it in sealed containers.

1801 Joseph-Marie Jacquard invents an automatic loom.

1824 Portland cement is patented by Joseph Aspdin.

1826 The gas stove is designed by James Sharp. Thomas Telford builds the first modern suspension bridge.

1844 John Gorrie builds a refrigerator.

1850s Oscar Levi Strauss invents modern jeans. Amelia Bloomer invents bloomers—pants for women.

1854 Elisha G. Otis develops a safe elevator.

1856 Henry Bessemer invents the Bessemer converter for making steel cheaply.

1867 The first modern cantilever bridge is built.

1869 Margarine is invented by Hippolyte Megè-Mouriès as an alternative to butter.

1876 Alexander Graham Bell and Elisha Gray independently file patents for telephones. Karl von Linde designs the first domestic refrigerator.

1882 The first hydroelectric power plant is built at Appleton, Wisconsin.

1883–1885 The first skyscraper is constructed in Chicago.

1892 François Hennebique patents reinforced concrete.

1901 Hubert Cecil Booth designs the first vacuum cleaner.

1907 Alva J. Fisher invents an electric washing machine.

1924 Clarence Birdseye's company begins making frozen food.

1988 The Seikan Tunnel under the Tsugaru Strait connects Japan's Honshu and Hokkaido islands. It is the longest rail tunnel in the world at 33.5 miles (53.9 km).

2009 Burj Khalifa in Dubai is completed. Reaching more than half a mile (0.8 km) into the sky, it is the world's tallest building.

2011 Danyang-Kunshan Grand Bridge opens on the high-speed railroad between Beijing and Shanghai. The bridge is more than 100 miles (160 km) long.

2019 The Mjøstårnet in Brumunddal, Norway, becomes the world's tallest wooden building. Its construction produced 85 percent less carbon emissions than a conventional skyscraper would have created.

2022 The 1915 Canakkale Bridge in Turkey is opened in Turkey. It is the longest suspension bridge in the world at 6,637 feet (2,023 m).

GLOSSARY

ancestors The people who came before the people alive today.

ancient Greece A civilization that existed on the mainland and islands of modern-day Greece and Turkey between 2000 and 300 BCE.

architect A person who designs buildings. The word comes from the Greek words for "chief builder."

atrium An open space inside a building.

buttress A structure, usually made from brick or stone, that is built against a wall to support and reinforce it.

cement A fine powder that is added to sand and water to make concrete. As the mixture dries the cement binds the sand together into a solid rock-like substance. Cement is made by heating limestone rocks. The first cements were naturally heated volcanic ashes.

centrifugal Relating to a force that seems to pull an object outward as it rotates around an axis.

cistern A water storage tank, often buried underground to keep it cool.

composite A material that is made from a mixture of other substances. The overall material takes on the characteristics of its ingredients. Composite materials include plastics and wattle and daub.

compression A force that tries to squeeze something into less space. A column pressed down is in compression.

corrode When a substance reacts with the water and salts in the surroundings, becoming weaker and disintegrating. Rust is the result of iron corroding.

deck The horizontal section of a bridge that carries the road, rail track, or walkway.

engineering Using scientific knowledge to solve problems and construct machines, buildings, and structures suited for a particular job.

foundations The place where a building meets the ground. Foundations in soft ground must be buried deeply to provide enough support.

gravity A natural force that attracts two masses toward one another. Among its many effects, gravity draws objects toward Earth's surface and keeps the planets in orbit around the Sun.

Industrial Revolution A great change in social and economic organization brought about by the replacement of hand tools by machines and power tools, and the development of large-scale industrial production methods. The Industrial Revolution started in England around 1760 and spread to the rest of Europe and the United States.

keystone The stone at the top of an arch where the opposite curves meet.

lock A gated section on a canal or river that allows craft to travel up and down steep gradients.

medieval Of or from the Middle Ages.

microwave An electromagnetic wave with a wavelength between 1 mm and 300 mm. Microwaves are used for radar, communications, and cooking and heating foods. Microwaves are a type of radio wave, but have a shorter wavelength—hence the name microwave.

Middle Ages A period of European history from around A.D. 500 to about 1450.

migration When people or animals move in search of a new place to live, better suited to their needs.

mortar A sticky paste that is used to glue bricks and stones together. Basic mortar contains mud and clay, but later forms also include cement.

mosque A Muslim place of worship, often characterized by a tall tower, called a minaret, from which the faithful are called to prayer.

nomads People who, instead of having a fixed home, travel from place to place in search of fresh pastures and water for themselves and their animals.

pollution Adding a large amount of unwanted materials into the air, soil, lakes, or oceans. Natural processes cannot always remove or absorb this extra material, which may build up, causing a range of problems.

porous When a solid is full of small holes through which liquids can pass. Many rocks are porous, so water can flow through them.

pressurized To have air packed into it to above the levels of the atmosphere.

radiation The act of giving off radioactive particles, heat, or electromagnetic waves.

refrigerant A fluid used to transfer heat from one space to another. The refrigerant is usually pumped around a closed circuit, absorbing heat from the cooler space and releasing it in the warmer space.

reservoir an artificial lake that is created behind a dam.

Rome The ancient civilization that began in the Italian city of Rome around 700 B.C. and had established a vast empire around the Mediterranean Sea by 200 A.D. The Romans are noted for being the first to bring law and order to Europe and for their great works of engineering.